RECENT TRENDS IN BIOTECHNOLOGY SERIES

RECENT TRENDS IN BIOTECHNOLOGY, VOLUME 1

Recent Trends in Biotechnology Series

Recent Trends in Biotechnology, Volume I
M. P. Singh, Anju Agrawal and Bechan Sharma (Editors)
2009. ISBN: 978-1-60876-148-7

RECENT TRENDS IN BIOTECHNOLOGY SERIES

RECENT TRENDS IN BIOTECHNOLOGY, VOLUME 1

M.P. SINGH
ANJU AGRAWAL
AND
BECHAN SHARMA
EDITORS

Nova Science Publishers, Inc.
New York

For permission to use material from this book please contact us:
Telephone 631-231-7269; Fax 631-231-8175
Web Site: http://www.novapublishers.com

LIBRARY OF CONGRESS CATALOGING-IN-PUBLICATION DATA

Recent trends in biotechnology / editors, M.P. Singh, Anju Agrawal, and Bechan Sharma.
p. ; cm.
Includes bibliographical references and index.
ISBN 978-1-60876-148-7 (hardcover)
1. Genetic engineering. 2. Biotechnology. I. Singh, M. P., Ph. D. II. Agrawal, Anju. III. Sharma, Bechan.
[DNLM: 1. Genetic Engineering--methods. 2. Biomedical Research. 3. Biosensing Techniques. 4. Biotechnology--trends. 5. DNA Fingerprinting. 6. Stem Cells. QU 450 R2956 2009]
TP248.6.R437 2009
660.6--dc22

2009031033

Published by Nova Science Publishers, Inc. ✢ New York

CONTENTS

PREFACE

Biotechnology has emerged as an area of activity having tremendous impact on virtually all domains of human welfare and holds a lot of promises and surprises in some of the techniques for the present and future. Spread over fourteen chapters, the book gives an overview of biotechnological means to convert hazardous wastes into useful products, then moves to an important area of biotechnology i.e. proteomics. It touches upon the latest research and developments in diverse areas like bio-informatics, DNA bio-sensors, DNA fingerprinting, mushroom biotechnology, stem cells, bioethics, in vivo induced antigen technology, immobilization technology, production of 7- aminocephalosporanic acid, therapeutic application of thymic peptides and cyanobacteria as a source of valuable products.

Chapter 1 - Management of voluminous quantities of hazardous wastes by physical, chemical and physicochemical means has proved to be inefficient, uneconomical and unecofriendly. On the other hand biotechnology which uses the biological processes has great potential to degrade the wastes into products that are part of naturally occurring cycles. Biotechnology includes a range of technologies which are applied in the purification of polluted air, treatment of wastewater and cleaning of contaminated soil. The bioremediation and phytoremediation techniques which use the microbial organisms and plants, respectively to detoxify hazardous wastes has received global attention as effective biotechnological approach to clean up the environment. Recombinant DNA technique has made it possible to create microorganisms with desired genes and subsequently use of these microorganisms in detoxification of hazardous wastes.

Chapter 2 - Proteomics involves the systemic study of proteins in order to provide a comprehensive view of the structure, function and regulation of biological systems. Advances in instrumentation and methodologies have fueled an expansion of the scope of biological studies from simple biochemical analysis of single proteins to measurement of complex protein mixtures. Proteomics is rapidly becoming an essential component of biological research. Coupled with advances in bioinformatics, this approach to comprehensively describing biological systems will undoubtedly have a major impact on our understanding of the phenotypes of both normal and diseased cells. Initially proteomics focused on the generation of protein maps using two-dimensional polyacrylamide gel electrophoresis. The field has since expanded to include not only protein expression profiling, but also the analysis of post-translational modifications and protein-protein interactions. However, mass spectroscopy has been incorporated to increase sensitivity, specificity and to provide results in a high throughput format. The study of protein-protein interactions has been revolutionized by the development of protein microarrays.

Chapter 3 - The development of databases and the use of bioinformatics have not only made science interesting but also easier to carry out new researches using the existing biological data. Enormous amount of the available data could be sorted out with ease as per the demand of the topic of the research using various algorithms, machine learning approaches and information retrieval systems. Genomics and Proteomics are an area of research which have still many mysteries unsolved and here the approaches of bioinformatics has made the tedious work much more simpler by providing numerous popular tools like BLAST and FASTA, ClustalW, PSI-Blast etc. to sort out the aggregate features from the completely sequenced genomic data. New methodologies are being developed to identify distant homologs, presence and absence of a protein in any sequenced genome, prediction of gene functions and many more. The role of bioinformatics in determining the complex structure of proteins, *de novo* modeling of loops in protein structure is worth mentioning. The revolution of bioinformatics in drug designing has given the hope to conquer the deadly diseases. The development of novel enzyme inhibitors like HIV-1 protease has now become possible. A number of programs for prediction of various ADME (Adsorption, distribution, metabolism and excretion) properties of hit molecules are successfully designed. Various computer-assisted drug discovery programs were developed for a wide range of applications including rigid body protein ligand docking, flexible docking and protein-protein interactions.

Chapter 4 - DNA based biosensors have recently gained much importance for detection of target genes responsible for diseases, in food industry, environment and in agriculture. Different types of DNA based biosensor, their advantages and basic principle of operating system has been described in this chapter. DNA biosensor provides fast, simple, economical, sensitive and selective detection of target genes by hybridization with specific probe. Various new strategies for DNA based biosensors have been described along with recent trends and challenges in development of technology. Electrochemical biosensor has more advantages due to the hybridization reaction on electrode surface or in solution in the presence of redox indicators. PCR free DNA biochip is emerging **as** new tools in the field of diagnosis.

Chapter 5 - DNA fingerprinting is a very powerful technology which is used by forensic scientists to establish identity. Such DNA-based technologies can help in the fight against crime. This technology has experienced many advancements occurring with the passage of time. Variable number tandem repeats (VNTR) are abundant in our genome and these regions are analyzed in DNA profiling.DNA testing is a matter of serious concern as it involves many ethical issues. DNA fingerprinting can be a boon to our society in different walks of our life if used efficiently.

Chapter 6 - Mushroom biotechnology is concerned with all aspects of the application of mushroom i.e. domestication of wild variety, enhanced production of nutriceuticals, bioconversion of lignocellulosic wastes into useful products, biodegradation of recalcitrant substances and bioremediation of xenobiotics. Molecular tools are being used for the improvement of mushroom strains which in turn can act as an effective weapon against malnutrition, many diseases and environmental pollution. Gene transfer using protoplast has been developed to break down the barrier of gene imposed by conventional breeding systems. Protoplast technology provides a highly feasible technique for interspecific and intergeneric hybridization. The versatile enzymatic systems of some of the mushrooms, which attack lignin and have low substrate specificity, are exploited to degrade many toxic and recalcitrant pollutants.

Chapter 7 - Treatment of tissue degenerative diseases like neurodegenerative disease, type one diabetes mellitus, spinal cord injury, chronic liver disorder, etc. and understanding various aspects of such diseases have remained as major challenges in the biomedical sciences. Isolation of embryonic stem (ES) cells opens up the possibility to meet these chanllanges although ethical and immunological issues in using these cells pose serious restrictions. Recent progress in developing methods to efficiently dedifferentiate cells from adult body into ES cells, known as induced pluripotent stem (iPS) cells, offers a promising and viable alternative. Use of iPS cells also bypasses ethical as well as immunological issues. IPS cell therapy, therefore, is being percied as the autogenic cell source for treatment of various tissue degenerative diseases.

Chapter 8 - Stem cells are cells found in most of the multi-cellular organisms. They are characterized by the ability to renew themselves through mitotic cell division and differentiating into a diverse range of specialized cell types. The prospect of this technology is immense despite the fact that research in the stem cell segment has been exhaustively discussed and criticized at numerous political, ethical and public forums in several countries. It has raised a hope among the scientist community about the possible use of stem cells as a tool not only to replace parts for failing organs but also to treat certain debilitating diseases like diabetes and Parkinson's disease. Most cells within an animal or human body are committed in fulfilling a specific function within the body. In contrast, stem cells are unique and important set of cells that are not specialized. These cells retain the ability to differentiate into many or all of the different cell types in the body and thereby play a critical role in repairing organs and body tissues throughout life. They can therapeutically divide without limit to replenish other cells as long as the person or animal is alive. When a stem cell divides, each new cell has the potential to either remain a stem cell or become another type of cell with a more specialized function, such as a muscle cell, red blood cell, or brain cell. Apart from application of stem cells in regenerative therapy, it also helps in understanding the early human development and the pathophysiology of genetic diseases at the cellular level. However, there are some technical problems and ethical concerns that need to be addressed before applying stem cells into clinical practice. The present chapter describes some salient features, current scenario and future prospects / retrospect of clinical applications of stem cells.

Chapter 9 - There are some common tenets to guide our actions and these are accepted as paradigm of morality in almost all human societies irrespective of differences in creed and culture. But in specific cases often unanimity is lacking. One basic criterion of morality is the acceptance of responsibility by the decision taker. New researches in the field of medical sciences are creating new problems which were hitherto not faced. New techniques in biotechnology have paved way for transplantation for kidney, heart and liver successfully. *In vitro* fertilization, surrogate mothership, researches in plelipotent stem cells which are capable of cloning are creating moral dilemma for the scientists. These techniques have immense capacity for the wellness of human beings. But there is every possibility of their misuse. Should researches in biotechnology be allowed unconditionally or there be some guiding principles? If the technique of cloning is perfected what will be the fate of the society which is based on conjugal love and trust? Should abortion and euthanasia be allowed? These questions need a clear answer. In our view no such laws can be framed which can guide scientists in each and every case. Each case is specific and requires particular attention. There should be some norms for individual decision taker which are given in spiritual book of Hindu 'Shrimad Bhagavata Geeta'. With the help of this technique of decision making specific cases can be dealt with.

Chapter 10 - IVIAT, abbreviated for *in vivo* induced antigen technology, is one of the recently invented techniques for identification of genes of a disease causing pathogen specifically induced upon infection of host. In simpler words these genes are switched on *in vivo* but remain switched off while growing *in vitro* outside the host. Since these genes are transcribed in the host environment, it is assumed that they may be essential for infection process, survival in the host environment and progression of disease. Identification of *in vivo* specific genes may thus be important in development of chemotherapeutics, vaccines and diagnostics. IVIAT offers several advantages over other techniques developed to select out genes induced *in vivo*. IVIAT circumvents the use of animal models and can be applied to detect temporal expression of antigens through infection process.

Chapter 11 - The immobilization of enzyme and cells is a practical method when bioprocess employing biocatalysts are considered. At present, the application of immobilized biocatalyst and cells in bioprocessing include production of energy, drugs, and enzymes, valuable proteins, specialty products etc. pollutant treatment, biosensors with high sensitivity and specificity through various techniques of immobilization protocol, numerous numbers of matrix materials were tried. Recent advancement in sophisticated instrumentation and recombinant DNA technology added new dimension to this technology not only considering the biological, biochemical aspects, but also of economical factors engineering aspects such as mass transfer resistance and diffusional limitations when dealing with this technology at commercial scale.

Chapter 12 - Cephalosporin acylases are industrially important enzymes, which hydrolyze cephalosporins to 7-aminocephalosporanic acid (7-ACA), a key intermediate required for the production of most of the clinically used cephalosporin derivatives i.e. semisynthetic cephalosporins. Cephalosporin-C (CPC), an important β-lactam antibiotic in the international market is exclusively produced by aerobic fermentation as such exhibits negligible antimicrobial activity but substitutions on C_3 and C_7 positions of β-lactam ring along with other structures generate semisynthetic cephalosporins with diversified antimicrobial activity e.g. cefazolin, cefotoxime, cephamandole and cephaclor etc. The production of 7-ACA and CPC was extensively reviewed in the chapter with some of our findings using both free and immobilized microbial cells.

Chapter 13 - Thymus, a gland responsible for production of T lymphocyte has been demonstrated to be a source of a variety of short peptide hormones collectively designated as thymic peptides. These polypeptides have gained increasing clinical therapeutic applications in a wide variety of patho-physiological anomalies cancer, neuroendocrine disorders, infections, senescence and autoimmune diseases. The thymic peptides depending on their chemical composition have been categorized into different families among which thymosins are the most investigated ones. Thymosin-α and –β have been shown to participate in immunomodulation and tumor progression. We have demonstrated from studies in our laboratory that thymosinα1 can reverse tumor induced immunosuppression for therapeutic benefit of the tumor bearing host. This chapter provides a brief review of current state of knowledge about the biological functions of thymic peptides and their potential for clinical applications and possible future directions for their therapeutic optimization.

Chapter 14 - Cyanobacteria, a small group of photosynthetic prokaryotes cosmopolitan in distribution have long been used as a model system to observe various physiological and biochemical processes. A little attention has so far been paid to the cyanobacterial secondary metabolites. The secondary metabolites not needed for normal growth exhibit antiviral, antibacterial, antifungal, antitumor, herbicidal, algicidal, antimalarial, inflammatory activities and if exploited reasonably can be of immense help to humankind. *Spirulina* is in use as human food because of its high nutritional value, high protein content (70% of dry weight) and presence of all the vitamins. Pigments especially phycobillins are in high demand in cosmetics and as food coloring agent. Cyanobacterial species are efficient catalysts of solar hydrogen production through photolysis of water. They also produce exopolysaccharides and flocculants useful in waste water treatment and clarification of solar ponds. Due to N_2 – fixing ability, cyanobacteria free – living or in symbiosis are often used as biofertilizer and reclamatization of saline usar soil in flooded paddy and sugarcane fields (specially in tropical soils). Traditional varieties of rice acquire their nitrogen need through biological sources such as cyanobacteria. *Azolla-Anabaena* association which reduces acetylene at rates that are 4-8 times greater than those observed for free-living *Anabaena cylindrica* is an efficient green manure. Cyanobacteria having simple growth needs may be exploited for the production of recombinant compounds of medicinal and commercial value as well as for the construction of mutants efficient in N_2-fixation and ammonium liberation. Cyanobacteria, unlike bacteria have been neglected as a source of valuable commercial products. This chapter emphasizes a need of multidisciplinary approach to harness the beneficial aspects related to biotechnological processes.

In: Recent Trends in Biotechnology, Volume 1
Editors: M.P. Singh, A. Agrawal and B. Sharma
ISBN 978-1-60876-148-7

Chapter 1

BIOTECHNOLOGY IN HAZARDOUS WASTE MANAGEMENT

M. P. Singh*
Department of Biotechnology, VBS Purvanchal University
Jaunpur – 222001 (India)

ABSTRACT

Management of voluminous quantities of hazardous wastes by physical, chemical and physicochemical means has proved to be inefficient, uneconomical and unecofriendly. On the other hand biotechnology which uses the biological processes has great potential to degrade the wastes into products that are part of naturally occurring cycles. Biotechnology includes a range of technologies which are applied in the purification of polluted air, treatment of wastewater and cleaning of contaminated soil. The bioremediation and phytoremediation techniques which use the microbial organisms and plants, respectively to detoxify hazardous wastes has received global attention as effective biotechnological approach to clean up the environment. Recombinant DNA technique has made it possible to create microorganisms with desired genes and subsequently use of these microorganisms in detoxification of hazardous wastes.

INTRODUCTION

Huge quantities of hazardous wastes are released into environment by agricultural, industrial, commercial, municipal, military, households and other human activities. This is contaminating air, water and soil, posing serious threat to human health and adversely affecting flora and fauna. Earth's ecosphere is overburdened with these wastes and biodiversity on the planet is under alarmingly endangered rate of extinction.

* E-mail: mpsingh.16@gmail.com; mpsingh_07@yahoo.com

The toxic effect of hazardous chemicals was reported in 1962 when Rachel Carson in his book 'Silent Spring' described how the use of pesticides, DDT in particular, affected human's health and also destroyed wildlife to such an extent that the spring arrived without the song of the birds. Carson's report on the deleterious effect of DDT residues on bird population led to the banning of DDT in the United States. Later on observation of carcinogenic effect of pesticides was realized. It was followed by the observation of the mercury poisoning along Minamata Bay in Japan where hundreds of cases of paralysis and sensory loss was reported. In the late 1960s and mid 1970s thousands of people became victim of contamination of cooking oil with polychlorinated biphenyls (PCBs) in Japan and Taiwan, respectively. Exposure of the high concentration of the chemical resulted into miscarriage and congenital defects.

The first international initiative to discuss environmental problems was taken in the United Nations conference on Human Environment in 1972 at Stockholm. In 1983 the report on world Commission on Environment and Development (WCED), 'Our Common Future', emphasized the need for an integration of economic and ecological systems. Another major effort was the United Nations Conference on Environment and Development (UNCED) held in 1992 at Rio de Janeiro. In December 2000 at Stockholm Convention 122 countries participated and finalized a treaty to ban 12 chemicals, the so called "Dirty dozen" among them, 8 are pesticides (aldrin, chlordane, DDT, dieldrin, endrin, heptachlor, minex and toxaphene), 2 are industrial chemicals (PCBs and hexachlorobenzene) and two byproducts of combustion and industrial process (dioxins and furans).This treaty was the first ever global agreement to abolish a class of chemicals. Most of the dirty dozen have already been banned in industrialized countries, but were still used in India and Latin America [3,5].

Management of huge amount of hazardous wastes by physical, chemical and physicochemical means has proved to be inefficient, insufficient and uneconomical. Here the degraded intermediate product sometimes may be more toxic, and environmental problem may partly be replaced. Hence, there is an urgent need to exploit the biotechnological innovations in dealing with hazardous wastes management. The pace with which new techniques have been developed and applied in biological research is remarkable. This is ranging from traditional biotechnology (food fermentation, biological control etc.) to modern biotechnology based on recombinant DNA technology, bioinformatics, DNA microarray, bioprocess technology, immunoassay etc.

Biotechnological methods have the advantages that use microbial mineralization process which convert xenobiotics and other pollutants into inorganic end products that are part of natural recycling processes. A whole range of technologies have been developed under biotechnology to transform hazardous wastes into harmless substances, generate biodegradable material from renewable sources, develop eco-friendly and safe manufacturing and disposal process and detect contamination and bioremediation.

HAZARDOUS WASTES

As per the definition of UNEP (United Nations Environment Programme) hazardous wastes are wastes that exhibit one or more hazardous characteristics, such as being flammable, oxidizing, poisonous, infectious, corrosive, or ecotoxic (Basel Convention).

Hazardous wastes in this chapter are referred to the substances that are harmful to health of humans, other organisms and environment. The population explosion followed by anthropogenic activities such as rapid industrialization and urbanization has not only increased the various type of hazardous wastes in the environment to critical levels but also produced a wide range of previously unknown contaminants in the form of xenobiotics for which our society and ecosystem is unprepared. Some of these wastes are highly recalcitrant, toxic and having high bioaccumulation and biomagnifications properties. The hazardous wastes production in developed countries such as United States, the Netherlands and Japan stand at 180, 100 and 5.5 kg per capita per year, respectively [12]. The waste consists of nonmetals, metals, metalloids, inorganic and organic compounds including aliphatic, alicyclic and aromatic hydrocarbons together with organooxygen, organohalides, organophosphate, organosulfur, organonitrogen, polyaromatic and mixed type of hydrocarbons. According to Rajendran and Gunasekaran [5], among 87000 commercial chemicals, the United States Environmental Protection Agency (USEPA) has identified 53 chemicals as persistence bioaccumulative and toxic (PBT). Of these, 42 are polychlorinated organic chemicals and other 11 are metals. They are the major environmental pollutants that have led to serious surface and sub-surface soil contamination.

Classification

Based on the carbon contents the hazardous wastes can be classified into organic and inorganic types.

Organic Wastes

Organic wastes are characterized by the presence of carbon element having carbon to carbon bonds. If the wastes contain only carbon and hydrogen they are called hydrocarbons. If carbon atom is bonded to a metal, chemically it is organometallic. If the organic compounds have the carbon backbone arranged in branched or straight chain they are aliphatic. On the other hand compounds having molecular structure with benzene (C_6H_6) as the basic unit are the aromatic compounds. Organic wastes can be natural or synthetic. Lignocelluloses comprising three major groups of polymers cellulose, hemicellulose and lignin are the most abundant natural organic compound available in the biosphere [7]. Pulp and paper, textile and other industries release unused cellulose residues, hemicelluloses and lignin as wastes. An organic compound can be synthetic, that is synthesized from chemicals found in feed materials such as petroleum, coal, wood or cultures of molds or bacteria [5].

Organic gaseous pollutants include hydrocarbons and other compounds such as formaldehyde, acetone vapours, alcohols, organic acids, methyl isocyanate, chlorinated hydrocarbon etc.

Inorganic Chemicals

Inorganic chemicals usually do not contain carbon, although some of them, such as sodium bicarbonate and sodium carbonate do contain carbon. A variety of salts containing carbonate, bicarbonate, acetate, nitrate, nitrite, fluorite, chloride, bromide, iodide, sulphate and phosphate present beyond threshold limit in water become harmful to flora and fauna.

Heavy metal such as copper, lead, mercury, arsenic, cadmium and selenium are hazardous when they reach beyond a limit.

Besides organic and inorganic wastes there are compounds which may be organic or inorganic but foreign in nature to biological systems. Such compounds are known as xenobiotic compounds. These are often toxic to life and are also usually very hard for microorganisms to metabolize since they contain molecular arrangement that are normally not present in nature. The presence of artificial groups such a chloro, nitro or sulphate in many synthetic chemicals makes them resistance to decomposition, as they are no longer recognized by the degrading microbes.

Sources

There are various sources for generation and release of hazardous wastes in the ecosystem; some of these are depicted below:

Agriculture

Huge quantities of agricultural and agro-industrial wastes are generated annually globally. India produces over 200 million tons of food grains annually and around 355 million tons of crop residues, out of which share of rice is 85 million tons, and over 100 million tons of paddy straw is generated as a byproduct [13]. According to Tiwari and Pandey [13] 50% of total crop residues is burnt and or left in the field as such. Paddy stubble is 'leftover' lower (2-4") portion of paddy crop, which includes roots. Keeping in view its heavier, denser and more bulky nature as compared to paddy straw it constitutes 5-10% of entire crop residue, and is burnt in the field [8].

Besides, the use of chemicals, fertilizers and pesticides in agricultural practice discharge hazardous pollutants. Pesticides contain the chemicals to protect crop from insects (insecticides), weeds (herbicides), nematodes (nimaticides), mites (acaricides), rodents (rodenticides) and disease causing microorganisms (fungicides and bactericides).

Municipal Wastes

It dumps large amounts of wastes mainly from urban areas. The wastes mainly contain bags, polypropylene, bottles, tubes, medical refuse, rubber, electronic refuse and other materials.

Industrial Wastes

The industrial wastes are primarily released from chemical industries, pulp and paper industries, oil refineries, sugar factories, tanneries, textile, steel, fertilizers and pesticides industries, coal and mineral mining industries, distilleries, metal processing industries, drugs, glass, cements, petroleum and engineering industries. Besides, dye industries discharges a lot of wastes.

Laboratories and Hospital Wastes

Laboratories and hospital wastes are generated by the research activities, diagnosis and treatment of human beings and animals. These include hazardous chemicals, pathogens, disposals, cultures, discarded medicines, soil wastes etc. These are in the form of disposable syringes, polypropylene bags, bandages, body fluids, human excreta.

Mining

In mining top soil and sub-soil is removed. This leaves deep salts in the earth. Mining operation may leave soil polluted with toxic heavy metals.

Impact

The advent of industrial revolution along with changed life style of human beings has resulted in impairment of environments through the release of hazardous wastes. Our ecosystem fails to biodegrade and mineralize the suddenly released huge quantity of non-metals, metals, metalloids, inorganic and organic wastes. The various chemicals in the wastes undergo a number of processes in the atmosphere, hydrosphere and lithosphere, transform into more toxic compounds and leave impact at all levels be it local, regional or global (Figure 1). Photochemical reactions, ozone depletion, acid rain and global warming are outcome of dumping of wastes in the atmosphere. Addition of fertilizers, pesticides, toxic metals, polyhalogenated and other organic compounds degrade soil and water, thus, badly affect hydrosphere and lithosphere.

On Air

Among naturally occurring organic substances, lignocellulosic materials are most abundant and resistant to degradation. Lignin component of lignocellulose cross react with other chemicals and transform into more recalcitrant compounds. The production of voluminous quantities of lignocelluloses is inherent outcome of agrarian economy of India. Major part of these lignocellulosic materials is burnt in the field resulting in multifaceted hazards including oxygen deficient environment, respiratory diseases and poor visibility at night [9, 10].

Emission from vehicles, thermal power plants, various industries and refineries are hazardous in nature, polluting air and causing global warming as well as climate change. It may cause various health problems in human being i.e. irritation to the eyes, nose and throat, respiratory problem like bronchitis and pneumonia and also headaches, nausea and allergic reactions. In the long run chronic respiratory disease, lung cancer, heart disease and even damaged to the nerves, brain, liver and kidney can take place.

On Water

Our surface water and groundwater are being polluted rapidly due to chemical effluents released from industries and agrochemicals. The use of organic compounds i.e. pesticides, fungicides, bactericides, herbicides, insecticides, nematicides, rodenticide etc. to protect agricultural product are responsible for acute water pollution problems, because these hazardous chemicals ultimately find their way into the nearby waterbodies. A number of biocides i.e., DDT, BHC, Aldrin, Dialdrin, chlordane, and endosulphan etc. when reach in the waterbodies cause severe health hazard to man and animals. These are mutagenic and may cause cancer.

Most of the conventional pesticides used in agriculture contain non-specific poisons that kill a broad spectrum of insects and pests, affecting other members of the local biosphere. The problems are further aggravated by the movement of these toxic chemicals in the ecosystem followed by bioaccumulation.

Release of chemical compounds from industrial wastes, urban wastes, agricultural practice, chemical and metallic pollutants, radioactive pollutants, mining etc. degrades soil and badly affects fauna and flora.

Toxic effluents discharge from various dyeing industries adversely affects water bodies, soil fertility, aquatic organisms and ecosystem integrity. They pose toxicity (lethal effect, genotoxicity, mutagenicity and carcinogenicity) to aquatic organisms as well as animals [4, 11].

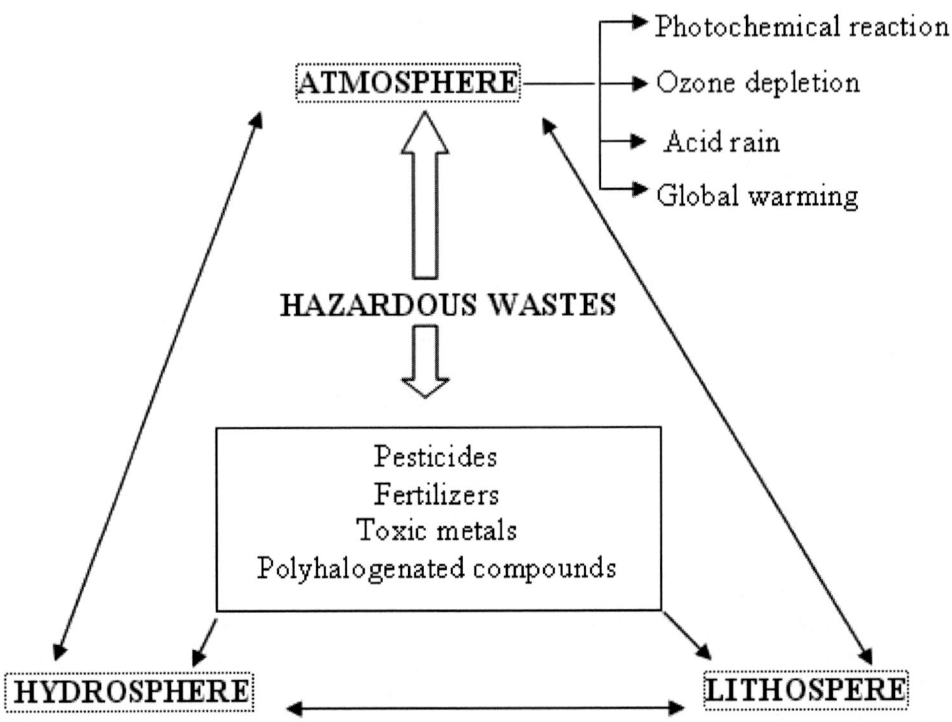

Figure 1. Impact of hazardous wastes on Ecosystem.

On Soil

Radioactive waste produced by nuclear reactors , weapon factories, explosion and nuclear devices pose a serious environmental risk since some of this waste remain radioactive for thousand of years. These wastes penetrate the soil and accumulate there, leading to land pollution. The waste cause chronic cellular damage to man and animals and can cause cancer and other serious ailments. Besides, waste generated from hospital and laboratories contain microorganisms like pathogens, bacteria, viruses and cause many diseases.

BIOTECHNOLOGY

Biotechnology has great promises for dealing with the hazardous wastes. Its application in this area with biosafety concerns towards environment and human health has increased decidedly in recent years. Biotechnology consists of gradient of technologies ranging from traditional biotechnology to modern biotechnology based on recombinant DNA technology, bioinformatics etc. It is used to detoxify the toxic hazardous wastes, cleanup the environment, generate the biodegradable materials and develop ecofriendly and safe manufacturing and disposal processes. Biotechnology relies on application of biological processes which includes microbial reaction to degrade and detoxify the hazardous wastes. Microorganisms involved in biological treatment process of hazardous wastes include bacteria, fungi, protozoa and alga.

Bioremediation Technique

The bioremediation technique involves the use of microbes to detoxify hazardous waste has received global attention as an effective biotechnological approach to cleanup the environment. Bioremediation, biodegradation, biotransformation, biodeterioration and cometabolosim are biological process but these terms are not always used with appropriate sensitivity to their subtle differences in meaning. According to Bennett *et al.* [1], in bioremediation biological systems are used to transform and/ or degrade toxic compounds or otherwise render them harmless. Bioremediation can involve indigenous microbial population with or without nutrient supplementation, or it can involve inoculation of exogenous organisms into the site. When exogenous organisms are added, the process is called 'bioaugmentation'. In either case, the goal is to disarm noxious chemicals without the formation of new toxins [1]. Biodegradation is biologically mediated breakdown of chemical compounds. When biodegradation is complete the process is called 'mineralization' i.e. the total breakdown of organic molecules into water, CO_2 and/or other inorganic end products. Biotransformation, a word often used synonymously with bioconversion usually refers to a single step in biochemical pathway, in which molecule (the precursor) is catalytically converted into a different molecule (the product). Biodeterioration is the breakdown of economically useful substances. Often the term is used narrowly to refer to the deterioration of substances that are normally resistant to biological attack such as metal, plastics, drugs, cosmetics, paintings, sculptures, wood products, electrical equipments, fuel and oils, and other valuable products [1, 6].

A wide range of bioremediation strategies has been developed to detoxify the hazardous wastes, consequently cleaning soil, water and air can be cleaned up. Bioremediation can be categorized into microbial bioremediation and phytoremediation. In microbial bioremediation, living microorganisms are used to convert complex toxic compounds into harmless byproducts of cellular metabolism such as CO_2 and H_2O. However, in phytoremediation, plants are used to remove contamination from the soil and water [12]. Microbial bioremediation mainly includes fungi and bacteria. Bacteria can be aerobic i.e. those which require oxygen to survive and anaerobic i.e. those that can survive without oxygen. Aerobic bioremediation is 10 to 100 times faster than anaerobic bioremediation.

Some of the examples of bacteria which are used for bioremediation are *Bacillus*, *Pseudomonas*, *Arthrobacteria*, *Flavobacterium*, *Deinococcus radiodurans* etc.

Mycoremediation

Fungal mediated remediation is referred to as mycoremediation. Fungal cells secrete extracellular enzymes which breakdown the hazardous wastes to derive energy for their growth and development. The effective degradation of lignocellulose is probably the single most important degradative event in the Earth's carbon cycle. Fungal degradative potential through extracellular enzymes are increasingly tried for degradation of non cellulosic wastes like plastic, petroleum hydrocarbon pollutants, dyes, pesticides and nutritional wastes.

Some of the fungal species which are increasing used in the degradation of hazardous wastes are *Phanerochaete chrysosporium*, *Pleurotus flordia*, *Trametes hirsute*, *Ceriporiopsis subermispora* etc. Presently *Pleurotus* species are used in the degradation of synthetic dyes efficiently due to presence of versatile enzymatic system [11]

Many of the fungi can absorb Cd, Cu, Pb, Hg and Zn into their mycelium and spores. Sometimes the walls of dead fungi bind better than living ones. Systems using *Rhizopus arrhizus* have been developed for treating U and Th [1, 14].

Recombinant DNA Technique

Recombinant DNA technique involves manipulation of the genetic material of an organism to give an altered expression at desired level. It deals with identification and isolation of desired genes and then insertion of this gene of interest into other organism. The desired gene expresses in the organism in which it is incorporated. This technique is also known as genetic engineering. Microorganisms are genetically modified using the recombinant DNA technique to remediate the hazardous wastes.

An extremely radiant-resistant bacterium, *Deinococcus radiodurans,* is being genetically engineered with biodegradation gene to render it suitable for the treatment of mixed waste [12].

Bacteria have evolved defense mechanisms against predation, an example being *Bacillus thuringiensis* which during sporulation, forms intracellular crystalline bodies that contain an insecticidal protein called the δ-endotoxin. The activated protein is highly poisonous to insects, some 80,000 times more toxic than organophosphate insecticide and is relatively selective, different strains of the bacterium synthesizing proteins are effective against larvae of different groups of insects [2].

Maize and many other plants have been engineered to produce δ-endotoxin to deal with the pest. Cotton which is grown, commercially in over 88 countries, is majorly affected by the pests. To fight against these pests without the external application of pesticides, genetically modified (GM) cotton has been developed for resistance against insects by incorporating insecticidal proteins from *Bacillus thuringiensis.* This GM cotton called Bt cotton has found extensive acceptance world over.

Purification of Polluted Air

With the advancement in biotechnology biofiltration systems have been developed to clean polluted air by means of microbiological degradation processes. Biofiltration uses microorganisms fixed to a porous medium to breakdown pollutants present in the air stream. The microorganisms grow in a biofilm on the surface of a medium or are suspended in water phase surrounding the medium particle. The filter- bed medium consists of relatively inert substances (compost, peat etc.), which ensure large surface attachment area and additional nutrient supply. As the air passes through the bed, the contaminants in the air phase absorbs into the biofilm and onto the filter medium, where they are biodegraded [12]. Figure 2 shows the flow chart of biofilter system used for purification of air. Microorganisms found in the packing material of biofilters have the potential of degrading a large number of organic compounds. Combination of processes used in this system is absorption, adsorption, degradation and desorption of gas- phase contaminants. With the technical advancement the packing material has been improved, ageing of filter has been retarded which has assured an extended period of relatively high biological activity. Visscher *et al.* [15] reported the application of biofiltration system, in 1978 in the Netherlands at sewage treatment plants to process polluted air that caused odour problems. The field of application was later on extended to industry for the removal of smell and biodegradation of chemicals like alcohols, esters, ketones etc. and xenobiotic compounds.

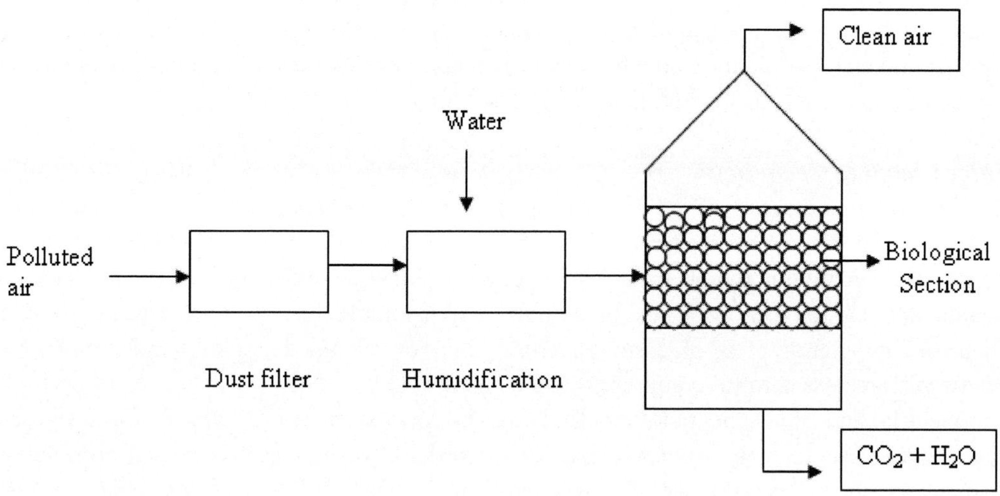

Figure 2. Flow chart of biofiltration system used for purification of air.

The overall effectiveness of biofilter is largely governed by the support medium, which include porosity, degree of compaction, water retention capabilities and the ability to host microbial populations. Critical biofilter operational and performance parameters include the microbial inoculums, medium pH, temperature and the medium moisture and nutrient content [12].

Because of the difficulties like accumulation of degradation products in biofilters, biotrickling filters are used. In the trickling filter, the gas flow moves concurrently and counter currently with a water phase through a bed of inert packing material with attached bilayers. Here, gas contaminants are absorbed in a free liquid phase prior to biodegradation by suspended or immobilized microorganisms.

Soil Cleaning Technique

The initial emphasis of thermal and extraction techniques of soil cleaning has recently been shifted to biological treatment techniques. The aim of the bilological soil treatment is to create fovourable environmental condition in the soil for improvement of the microbiological activity and consequently enhancement of the biodegradation of the contaminants. Favorable condition for the microbes mean roughly sufficient oxygen and nutrients supply, a pH ~ 7, and temperature between 25 and 30^{0}C, the moisture content and a good soil structure are also very important for biological activity.

The biological treatments for soil cleaning includes composting, land farming, *in situ* biorestoration and bioreactors. In composting microorganisms such as bacteria, fungi convert degradable organic wastes into humus like substance which is high in carbon and nitrogen content and is an excellent medium for growing plants. Composting ensure the recycling and enrich the soil. Besides being clean, inexpensive and safe, composting not only convert the hazardous organic wastes but also increase the water holding capacity of soil. On the other hand in simple landfarming method, the contaminated soil is spread over a sand layer with a drainage system to a depth of approximately 40 cm. Prior to this, the soil is protected by means of a plastic layer (aprroximately 0.5 mm of polyvinyl chloride). Landfarming is also carried out under so called a conditioned circumstances. This mean that during the cleanup process, a better control of parameters such as oxygen, water contents, and temperature is achieved [15]. *In situ* boidegradation stimulates the natural degradation processes in soil. Here ground water is pumped up and is used as medium of oxygen, nutrients, and eventually detergents and organisms to stimulates boidegradation of the contaminants.

Water Cleaning Technique

Biological treatment method employing both aerobic and anaerobic microorganisms are used to detoxify the toxic organic and other substances present in water since this can remove most of biologically removable organics and colour. The liquid effluents contain thousands of organic and inorganic substances in soluble and insoluble forms in industrial effluents. Controlled population of aerobic and anaerobic microorganisms degrade complex chemicals into simple and less harmful compounds and elements [12].

Bacteria and fungi degrade most of the pollutants of water. White rot fungi are increasingly used recently to biodegrade the hazardous wastes because of the presence of versatile enzymatic machinery. These fungi have batteries of enzymes like laccase, manganese peroxidase (MnP), lignin peroxidase (LiP), veratryl peroxidase and glyoxal oxidase (GLox) which have no or low substrate specifity and can attack variety of compounds resembling lignin structure. This property is exploited for faster and complete degradation of recalcitrant and xenobiotic compounds. The detoxification of wastewater by biological method is advantageous over physical and chemical method.

CONCLUSION

The issues of population, pollution and production are linked. Population explosion has led to release of huge quantities of hazardoud wastes into environment by agricultural, industrial and other human activities. This is posing serious threat to human health and adversely affecting flora and fauna. There is an urgent need to save the earth's overburdened ecosphere for the survival of human beings and other organisms. Biotechnology offers the means to convert the wastes into food, feed, energy and other products alongwith clean and green environment. It has advantages over physical, chemical and pysicochemical means of management of hazadous wastes since it provides ecofriendly safe disposal processes.

ACKNOWLEDGMENT

The author is thankful to Prof. R. C. Saraswat, Vice-Chancellor, VBS Purvanchal University, Jaunpur (India) for encouraging qualitative research in the University Campus and affliating colleges.

REFERENCES

[1] Bennett, J.W., Wunch, K.G. and Faison, B.D. 2002. Use of fungi in biodegradation. In: C.J. Hurst (ed.), *Manual of Environmental Microbiology*, ASM Press, Wshington, DC, pp 960-971.

[2] Brown, T.A. 2001.Gene cloning and analysis in agriculture. In: *Gene Cloning.* Fourth Edition, Blackwell Publishing Company, U.K., pp. 313-328.

[3] Hill, M.K. 2004. Persistent, bioaccumulative and toxic pollutants. In: *Understanding Environmental Pollution.* Second edition, Cambridge University Press, U.K., pp 338-349.

[4] Puvaneswari, N., Muthukrishnan, J. and Gunasekaran, P. 2006. Toxicity assessment of microbial degradation of azo dyes. *Indian J. Experimental Biology,* **44**:618-626.

[5] Rajendran, P. and Gunasekaran, P. 2006. *Microbial bioremediation.* MJP publisher, Chennai., pp 1-282.

[6] Rose, A.H. 1981. *Microbial Biodeterioration*, vol. 6, *Economic Microbiology*, Academic Press, Ltd., London, U.K.

[7] Singh, M.P. 2000. Biodegradation of lignocellulosic waste through cultivation of *Pleurotus sajor-caju*. In: Van Griensven (ed.), *Science and Cultivation of Edible Fungi.* Balkema, Rotterdam, pp 517-521.

[8] Singh, M.P. and Gautam, N.C. 2004. An overview of lignocellulose biotechnology. In: N.C. Gautam and M.P. Singh, (eds.), *Recent Advances in Biotechnology*. Shree Publishers and Distributors, New Delhi, pp. 3-20.

[9] Singh, M.P. and Sharma, R. 2002. *Pleurotus florida* Eger- an effective biodegrader of steam sterilized lignocellulosic waste. *Poll. Res.* **21**(1): 63-67.

[10] Singh, M.P., Pandey, V.K., Pandey, A.K., Srivastava, A.K., Vishwakarma, N.K., and Singh, V.K. 2008. Production of Xylanase by white rot fungi on wheat straw. *Asian Jr. of Microbiol. Biotech. Env. Sc.* **10**(4) : 859-862.

[11] Singh, M.P., Rastogi, P.C., Srivastava, A.K. and Vishwakarma, N.K. 2008. Decolorization of azo dyes by white rot fungi *Pleurotus* species. *Poll. Res.* **27**(3): 365-369.

[12] Thakur, I.S. 2006. *Environmental biotechnology basic concept and applications*. I.K. International Pvt. Ltd., New Delhi, pp 1-472.

[13] Tiwari, R.P. and Pandey, M. 2002. Sizable income generating venture. In: Ravi (ed.), *Hindu Survey of Indian Agriculture*, pp.165.

[14] Treen- Seares, M.E., Martin, S.M. and Volesky, B.1984. Propagation of *Rhizopus javanicus* biosorbent. *Appl. Environ. Microbiol.* **48**:137-141.

[15] Visscher, K., Brinkman, J. and Soczo, E.R. 1990. Biotechnology in hazardous wastes management in the Netherlands, In: D. Kamely, A. Chakrabarty, and G.S. Omenn (Editors), *Biotechnology and Biodegradation, Advances in applied biotechnology* series Vol.4,. Portfoilo Publishing Company, the Woodlads, Texas, pp. 389-403.

In: Recent Trends in Biotechnology, Volume 1
Editors: M.P. Singh, A. Agrawal and B. Sharma
ISBN 978-1-60876-148-7

Chapter 2

PROTEOMICS

Anju Agrawal[1] and Bechan Sharma[2]
[1]Department of Zoology, SNSen BVPG College
CSJM University Kanpur, India
[2]Department of Biochemistry, University of Allahabad
India

ABSTRACT

Proteomics involves the systemic study of proteins in order to provide a comprehensive view of the structure, function and regulation of biological systems. Advances in instrumentation and methodologies have fueled an expansion of the scope of biological studies from simple biochemical analysis of single proteins to measurement of complex protein mixtures. Proteomics is rapidly becoming an essential component of biological research. Coupled with advances in bioinformatics, this approach to comprehensively describing biological systems will undoubtedly have a major impact on our understanding of the phenotypes of both normal and diseased cells. Initially proteomics focused on the generation of protein maps using two-dimensional polyacrylamide gel electrophoresis. The field has since expanded to include not only protein expression profiling, but also the analysis of post-translational modifications and protein-protein interactions. However, mass spectroscopy has been incorporated to increase sensitivity, specificity and to provide results in a high throughput format. The study of protein-protein interactions has been revolutionized by the development of protein microarrays.

INTRODUCTION

Proteomics is a new science that focuses on the large-scale study of proteins, particularly their structure and function [1, 2], localizations, interactions and other factors. Proteins in turn

[1] E-mail: anjuaa@rediffmail.com
[2] Corresponding author E-mail: sharmabi@yahoo.com

are vital parts of living organisms, as they are the main components of the physiological metabolic pathways of cells. The term proteomics was first coined in 1997 [3] to make an analogy with genomics, the study of the genomes of organisms. The word `Proteome` is a blend of protein and genome and was coined by Prof. Marc Wilkins in 1994 while working on the concept as a PhD student [4, 5].

The proteome is the entire complement of proteins [4], including the modifications made to a particular set of proteins, produced by an organ or system. This will vary with time and distinct requirements, or stresses, that a cell or organism undergoes. The word "proteome" is derived from PROTEins expressed by a genOME, and it refers to all the proteins produced by an organism, much like the genome is the entire set of genes. The human body may contain more than 2 million different proteins, each having different functions. As the proteins are main component of the physiological pathways of any cell, they serve vital functions in the body such as: (i) catalyzing various biochemical reactions e.g. enzymes, (ii) acting as messengers for example, neurotransmitters, (iii) acting as control elements that regulate cell reproduction, (iv) influencing growth and development of various tissues e.g. trophic factors, (v) transporting oxygen in the blood e.g. hemoglobin and (vi) defending the body against diseases e.g. antibodies.

Proteins are fairly large molecules made up of chain(s) of amino acids, arranged in different combinations to make tens and thousands of proteins, each with a unique genetically defined sequence that determines specific shape and function of a protein. In addition each protein can undergo a variety of post translational modifications that further influence its shape and function. Researchers and scientists are working on developing a map of the human proteome–much like that of human genome–that identifies novel protein families, protein interaction and signaling pathways. The term proteomics encompasses a broad set of disciplines aimed at understanding and monitoring proteins. This includes work correlating genetic sequences with 3D protein structure and sequence, development of protein profiling techniques, investigation of protein–protein interaction, promises to discover unanticipated targets for drug design by determining the thousand of unidentified proteins expected to be found in human genome. It would also provide experimental data to improve computer-modeling programs through protein structure from DNA sequence. This may help improve the interpretation of information contained in the genome. The explosive growth of this field is driven by multiple forces such as genomics and its relative of several new proteins; powerful protein technologies such as newly developed mass spectrometry approaches, global two hybrid technologies, spin offs from DNA arrays; innovative computational tools/ methods to process, analyze and interpret prodigious amounts of data.

Unlike the genome which is relatively static, the proteome changes constantly in response to tens and thousands of intra and extracellular environmental signals. The proteome varies with health or disease, nature of the each tissue, stage of cell development and effects of drug treatments, biochemical interactions with genome or environment. One organism has radically different protein expression in different parts of its body, in different stages of its life cycle and in different environmental conditions. As such the proteome often is defined as the "Proteins in one sample (tissue, organism, cell culture) at a time." In many ways proteomics runs parallel to genomics. Genomics starts with the gene and make inferences about its products (Proteins) whereas proteomics begins with the functionally modified protein and works back to the gene responsible for its production. The sequencing of the human genome has increased interest in proteomics because while DNA sequence information provides a static snapshot of the various ways in which the cell might use its proteins, the life of a cell is a dynamic process. This data set holds great new promises for proteomic applications in science, medicine and most notably – Pharmaceuticals.

Proteomics is often considered the next step in the study of biological systems, after genomics. It is much more complicated than genomics, mostly because while an organism genome is more or less constant, the proteome differs from cell to cell and from time to time. This is because distinct genes are expressed in distinct cell types, meaning that even the basic set of proteins, which are produced in a cell, needs to be determined. In the past this was done by mRNA analysis, this was found not to correlate with protein content [6, 7]. It is now known that mRNA is not always translated into protein [8] and the amount of protein produced for a given amount of mRNA depends on the gene it is transcribed from and on the current physiological state of cell. Proteomics confirms the presence of proteins and provides a direct measure of the quantity present.

It is worth mentioning that in the pre-proteomic era, thousand of proteins were exquisitely characterized –those in metabolic and signaling pathways; in the replication, transcription and translation machinery, in secretary and in other cellular complexes. These functional assignments came from the efforts to understand specific cellular processes, with three major factors fueling progress in the last two decades: (i) a confluence of geneticists, cell biologists, biochemists and structural biologists arrived to tackle the same problems from different directions, (ii) the extraordinary conversation of fundamental cellular mechanism enabled insights garnered from the study of one organism to be immediately applied to all others–what is true for the walrus is true for the carpenter and (iii) developments in technology–the new standard tools for molecular biology such as DNA sequencing, recombinant DNA and the polymerase chain reaction (PCR)- sparked new experimental strategies.

With the emergence of proteomics additional proteins are now pouring into join those already implicated in some process or the other. This extra information however derives not from continued small-scale analysis of biological activity, but from much larger and more systemic studies. Proteomics like its precursor genomics thus represents the emergence of a new way of doing research that is not dependent on the testing of specific models of cellular behavior. This style of science obviously does not replace, but rather will increasingly operate in tandem with, traditional biological research methods. A big boost has come from recent advances in mass spectroscopy that allow the rapid identification of proteins separated in 2D gel or by chromatography. The mass spectrometer measures the masses of peptides, which are then compared to the predicted masses of peptides from silico digestion of sequences in genomic databases. Although the unambiguous identification of a protein cannot always be derived from the masses of few of its peptides, in the tandem mass spectrometer, peptide ions from the first mass spectrometer run are fragmented and identified in a second run to yield the more valuable commodity of a peptide sequence. A single peptide sequence usually identifies a protein.

Advances in automation increased sensitivity and throughput combined with improved biochemical fractionations and the availability of vastly expanded databases have extended the application of mass spectrometry to even bigger jobs. For example, megadalton protein complexes can be purified, often with a single tagged component, and their constituents can be identified after gel electrophoresis. Such analysis has been performed on among other complexes, the human spliceosome. By passing even gel separation, the direct analysis of protein complexes identifies components of heterogeneous protein mixtures often using 1D or 2D chromatography for fractionation before analysis of mass spectroscopy. Application of this procedure to a whole cell yeast lysate identified 189 proteins and more recently 1484 proteins including integral membrane proteins and those of low abundance in the cell.

It is important to mention that any particular protein may go through a wide variety of alterations, which may have critical effects to its function. For example, during cell signaling many enzymes and structural proteins can undergo phosphorylation. The addition of a phosphate to particular amino acid most commonly serine and threonine [9] mediated by serine / threonine kinases or more rarely, tyrosine mediated by tyrosine kinases causes a protein to become a target for binding or interacting with a distinct set of other proteins that recognize the phosphorylated domain. As protein phosphorylation is one of the most studied protein modifications, many "Proteomic" efforts are geared in determining the set of phosphorylated protein in a particular cell or tissue type under particular circumstances. This alerts the scientists to the signaling pathways that may be active in that instance.

Ubiquitin is a small protein that can be affixed to certain protein substrate by enzymes called E3 ubiquitin ligases. The determination of polyubiquitinated proteins can be helpful in understanding how protein pathways are regulated. This is therefore an additional legitimate "Proteomic" study. Similarly, once it is determined what substrates are ubiquitinated by each ligase, determination of the set of ligases expressed in a particular cell type would be helpful. In addition to phosphorylation and ubiquitination, protein can be subjected to methylation, acetylating, glycosylation, oxidation, nitrosylation etc. Some proteins undergo all of these modifications, which nicely illustrate the potential complexity one has to deal with when studying protein structure and function.

In order to attain a 3D functional conformation of the protein structure and the biological activity, many nascent polypeptides undergo posttranslational modification before or after the folding process. This is especially true in eukaryotic cells where most modifications take place in two important subcellular organelles such as endoplasmic reticulum and golgi apparatus. The modifications include proteolytic cleavage; formation of disulfide bonds; addition of phosphoryl, methyl, acetyl or other chemical characteristic groups into certain amino acid residues; or attachment of oligosaccharides or prosthetic groups to create mature proteins. Posttranslational modifications have great impact on protein function by altering the size, hydrophobicity and overall conformation of the proteins. The modifications can directly influence protein-protein interactions and distribution of proteins to different subcellular locations. It is therefore important to use bioinformatics tools to predict sites for posttranslational modification. This often leads to many false positive identifications. One such example is the known consensus motif for protein phosphorylation, [ST]-X-[RK]. Such a short motif can be found multiple times in almost every protein sequence. Most of the predictions based on the sequence motif alone are likely to be wrong, producing very high rates of false-positives. Similar situations can be found in other predicted modification sites. One of the reasons for false predictions is that the neighboring environment of the modification sites is not considered.

To minimize the false positive results, a statistical learning process called support vector machine (SVM) can be used to increase the specificity of prediction. In this method, they are projected in a three dimensional space or even a multidimensional space. Automotif (http://automotif.bioinfo.pl/) is a webserver predicting protein sequence motifs using the SVM approach. In this process, the query sequence is chopped up into a number of overlapping fragments, which are fed into different kernels (similar to nodes). A hyperplane, which has been trained to recognize known protein sequence motifs, separates the kernels into different classes. Each separation is compared with known motif classes, most of which are related to posttranslational modifications. The best match is with a known class, which defines the functional motif; even if one is studying a particular cell type, that cell may make different sets of proteins at different times, or under different conditions. Furthermore, any protein can undergo a wide range of post-translational modifications. That is why a proteomics study can become quite complex very quickly, even if the object of the study is very restricted.

In understanding protein function, the key requirement is to learn to correlate the vast array of potential protein modifications to particular phenotypic settings, and then to determine if a particular posttranslational modification is required for functions to occur. There has been great progress in uncovering gene expression measures that are clinically informative, and suggest pathway association with disease. Nevertheless, two key challenges remain to translate these gene relationships to new diagnostic and ultimately to new therapies: (i) to establish a correlation between the prospective gene markers and their corresponding protein products, especially the subsets most relevant to clinical features of the disease and (ii) to distill most informative subset of genes to assayable biomarkers that can be measured from biological fluids using technologies common to laboratories (Fig 1).

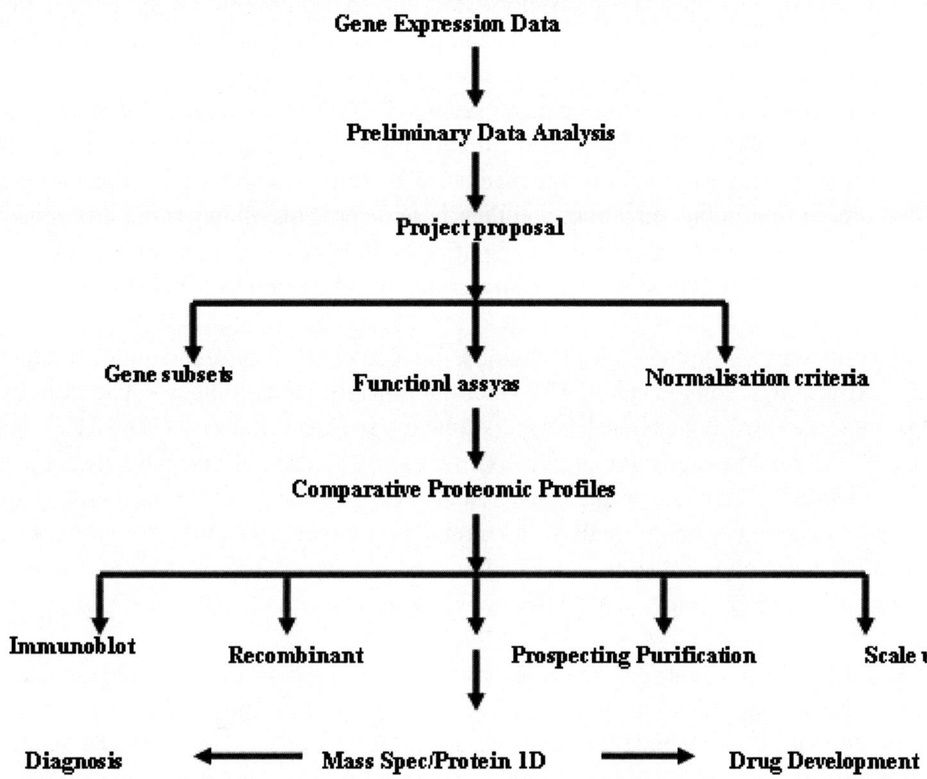

Figure 1. Flow chart showing different proteomic profiles and its applications.

In this context, ProFACT's services assist gene expression researcher challenges by providing: (i) a bioinformatics bridge to determine the most practical measurable protein attributes from a pre-selected gene panel and (ii) a wide assortment of proprietary and conventional tools and methods to generate proteomic datasets based on measurable protein attributes.

Using a multidimensional proteomic approach, ProFACT will identify, quantify and analyze predicted protein targets based on expression data.

Proteomics gives a much better understanding of an organism than genomics because (i) the level of transcription of a gene gives only a rough estimate of its level of expression into a protein. mRNA produced in abundance may be degraded rapidly or translated inefficiently, resulting in small amount of protein, (ii) most proteins experience post translational modifications that profoundly affect their activities, for example some proteins are not active until they become phosphorylated. Then methods such as phophoproteins and glycoproteomics are used to study post-translational modifications, (iii) many transcripts give rise to more than one protein, although alternative splicing or alternative post–translational modifications may occur, (iv) many proteins form complexes with other proteins or RNA molecules may exist and only function in the presence of these or other molecules, and (v) protein degradation rate plays an important role in protein content [10].

One way in which a particular protein can be studied is to develop an antibody, which is specific to that modification. For example, there are antibodies, which only recognize certain proteins when they are tyrosine–phosphorylated and there are antibodies specific to other modifications. These can be used to determine the set of proteins that have undergone the modification of interest for sugar modifications, such as glycosylation of proteins. Certain lectins have been discovered which bind to sugars. These too can be used. A more common way to determine posttranslational modification of interest is to subject a complex mixture of proteins to electrophoresis in "two dimensions, 2D" which simply means that proteins are electrophoresed first in one direction, and then in another. This allows small differences in a protein to be visualized by separating a modified protein from its unmodified form by this methodology, which is known as "two dimensional polyacrylamide gel electrophoresis (2D-PAGE)".

The commonly used proteomics techniques include (1) Protein separation techniques, (2) Protein extraction / solubilisation, (3) Cell fractionation and protein separation by one dimensional and two dimensional polyacrylamide gel electrophoresis (1D-PAGE and 2D-PAGE), (4) Chromatography for protein purification, (5) Protein detection techniques, (6) Two-dimensional gel electrophoresis (isoelectric focusing electrophoresis, capillary electrophoresis etc), (7) Image analysis for proteomics experiments, (8) Protein and peptide identification by peptide fingerprinting, amino acid sequence analysis, database search etc., (9) Characterization of protein post-translational modification and (10) Functional proteomics study.

The 2D-PAGE is a high-resolution technique that separates protein by charge and mass. The gel is run in one direction in a pH gradient and then in orthogonal dimension under a denaturing condition to separate proteins by molecular weights. This is followed by staining, usually silver staining, which is very sensitive, to reveal the position of all proteins .The result is a two dimensional gel map; each spot on the map corresponds to a single protein being expressed. The stained gel can be further scanned and digitized for image analysis (Figure 2). However, not all the protein can be separated by this method or stained properly. One of the challenges of this technique is the separation of membrane bound proteins, which are largely hydrophobic and not readily solublized. They tend to aggregate in the aqueous medium of two-dimensional gel. To overcome this problem, membrane protein can be fractionated using specialized protocols and then electrophoresed using optimized buffer containing zwitter ionic detergents. The subcellular fractionation can be carried out to separate nuclear, cytosol, cytoskeletal and other subcellular fractions to boost the concentrations of rare proteins and to reveal subcellular localizations of the proteins.

The gel image analysis is the next step that helps reveal differential global protein expression patterns. This analysis includes spot determination, quantitation and normalization. The image analysis software is used to measure the center, edges and densities of the spots. Comparing two dimensional gel images from various experiments sometimes pose a challenge because the gels, unlike DNA microarrays, may shrink or wrap. This requires the software programs to be able to stretch or maneuver one of the gels relative to the other to find a common geometry. When the reference spots are aligned properly, the rest of the spots can be compared automatically.

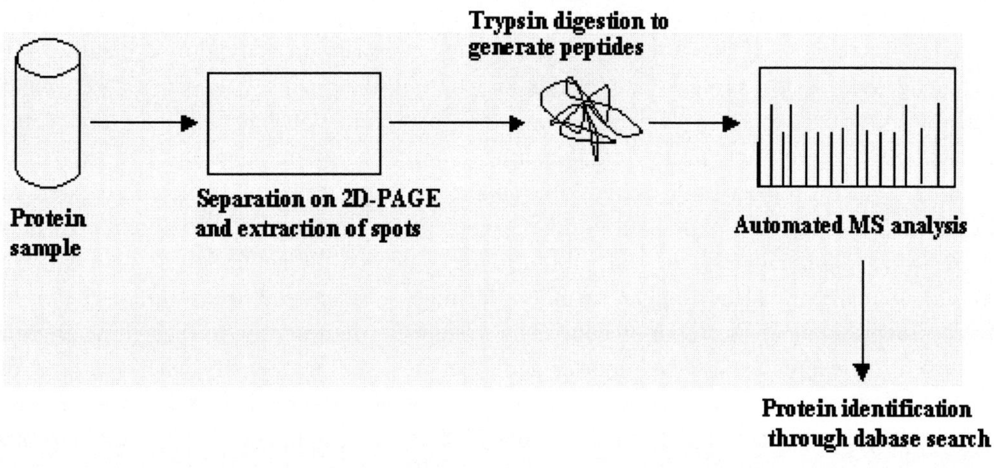

Figure 2. A schematic representation of the procedure for proteome characterization using 2D-PAGE and Mass Spectroscopy (MS).

Classically, antibodies to particular proteins or their modified forms have been used in biochemistry and cell biology studies. These are among the most common tools used by practising biologist today. For more quantitative determination of protein amounts, techniques such as ELISA can be used. For proteomic study, more recent techniques such as Matrix-assisted laser deposition/ionization (MALDI) or Matrix-assisted laser desorption/ionisation-time of flight mass spectrometry (MALDI-TOF MS) have been employed for rapid determination of proteins in particular mixtures.

MALDI-TOF MS is a relatively novel technique in which a co-precipitate of an UV-light absorbing matrix and a biomolecule is irradiated by a nanosecond laser pulse. Most of the laser energy is absorbed by the matrix, which prevents unwanted fragmentation of the biomolecule. The ionized biomolecules are accelerated in an elctric field and enter the flight tube. During the flight in this tube, different molecules are separated according to their mass to charge ratio and reach the detector at different times. In this way each molecule yields a distinct signal. The method is used for detection and characterization of biomolecules, such as proteins, peptides, oligosaccharides and oligonucleotides, with molecular masses between 400 and 350,000 Da. It is a very sensitive method, which allows the detection of low (10^{-15} to 10^{-18} mole) quantities of sample with an accuracy of 0.1 - 0.01 %. Protein identification by this technique has the advantage of short measuring time (few minutes) and negligible sample consumption (less than 1 pmol) together with additional information on microheterogeneity (e.g. glycosylation) and presence of by-products.

It is known that most proteins function in collaboration with other proteins. One purpose of using proteomics is to identify which proteins interact. This is especially useful in determining partners in cell signaling cascades. Several methods are available to probe protein-protein interactions. The traditional method for identifying protein-protein interactions in biological systems is a yeast two-hybrid analysis. One can use this technique for selection and can survey millions of different protein-protein interactions in yeasts. Only those yeasts that harbor a productive protein-protein interaction would grow. If they grow, they form colonies, which can be picked up and used for analysis. Some new methods include protein microarray, immunoaffinity chromatography, mass spectroscopy (MS) and experimental methods such as phage displays and computational methods.

The two-hybrid technique exploits eukaryotic transcription factors that consist of two modular domains: a DNA–binding domain and transcription-activation domain. In this technique, two fusion proteins are made. The first is the fusion between a DNA-binding domain and a bait protein. Typically, the bait is a protein you have in hand, and you are interested in determining protein-protein interactions that participates in. The second is a fusion between a transcription-activation domain and a prey protein, a potential interacting partner. Both fusions are expressed in the presence of a reporter gene. If bait and prey do not interact, the reporter gene is not turned on. If they interact, the activation domain and DNA binding domain are also brought together, forming a functional transcription factor that induces expression of the reporter gene. The system is often set up so that yeast growth is dependent on expression of the reporter gene-that is, yeast grows only when bait and prey interact. Since the basic two-hybrid technique was developed in the late 1980s, scientists have improved and modified it extensively. There are many examples of conditional protein-protein interactions, where, for example, binding of the two proteins to each other is dependent on phosphorylation of one of the proteins.

One can also uncover protein-protein interactions that are dependent on binding of a small-molecule ligand or the presence of a protein or RNA that forms a bridge between bait and prey. The original scope of the yeast two-hybrid screen has been broadened by these modifications. Erica A. Golemis and coworkers at Fox Chase Cancer Centre, Philadelphia, recently developed a dual bait variation of the yeast two hybrid system that analyses two independent protein-protein interactions simultaneously [11]. It is explained that the dual-bait system excels at distinguishing specific from nonspecific interactions. For instance, it can be used to identify two discrete sets of interacting proteins against an excessive background population of nonspecifically interacting proteins. Possibly, the most notable recent yeast two-hybrid study was used in the technique to create the first protein-protein interaction map of an entire organism [12]. About 957 possible protein-protein interactions have been detected involving 1,004 yeast proteins.

Protein microarray chips are conceptually similar to DNA Microarray chips and can be built to contain high-density grids with immobilized proteins for high throughput analysis.

The chip contains entire immobilized protein. However, they are not meant to be used to bind and quantitate complementary molecules as in DNA Microarrays. Instead, they are used for studying protein function by providing a solid support for assaying enzyme activity or protein-protein interaction, protein-DNA/RNA interaction or protein–ligand interaction in or all-against–all format. To make protein chips truly analog to DNA chips, the solid support has to contain specific or ligands that capture protein molecules by complementarities. A classical approach to this problem is to perform an immunoassay by using a spectrum of the antibodies against the whole proteome. The antibodies can be fixed on a solid support for assaying thousands of proteins simultaneously. However, a natural drawback of this approach is that natural antibodies are easily denatured and have a high tendency to cross react with non-specific antigens. In addition, the production of antibodies for every single protein by an organism is prohibitively expressive.

To overcome this hurdle a new technique is being developed that uses "protein scaffolds" to capture target molecules. The scaffolds are similar to antibodies but smaller, highly stable and specific in their binding of target protein. They can be made in a cell free system and attached with two fluorescence tags. This technique uses the principles of fluorescence resonance energy transfer (FRET) between two fluorescent dye molecules whose excitation and absorption spectra overlap. The efficiency of the energy transfer depends upon the distance of the two dyes. If one portion of tagged protein is involved in binding to a target protein, disruption of the excitation energy transfer between the dyes is such that it can be monitored on fluorescence spectra. A technology called Protein–Print (PP) is in early development, which is essentially a molecular imprinting method. Chemical monomers are used to coat target proteins, which are then allowed to polymerize. When polymerization is complete the target molecule is removed, a mould is formed that resembles the shape of the 0target protein. The moulds can than be used to capture like molecules with high specificity.

After separation of proteins by 2D-PAGE, they can be further identified and characterized using MS. In this procedure, the proteins from a two-dimensional gel system are first digested *in situ* with the protease (e.g. trypsin). Proteins spots of interest are excised from the gel. The proteolysis generates a unique pattern of peptide fragments of various molecular weights, which is termed as a peptide fingerprint. The fragments can be analyzed with MS, a high-resolution technique for determining molecular masses. Currently, in some cases instead of using MS, electrospray ionization MS (ESIMS) and MALDI are used. These two approaches only differ in ionization procedure. In MALDI-MS for example, the peptides are charged with positive ions and forced through an analyzing tube with a magnetic field. Peptides are analyzed in the gas phase. Since the smaller peptides are deflected more than larger ones in a magnetic field, the peptides fragments can be separated according to their molecular masses and charges. A detector generates a spectrum that displays ion intensity as a function of mass-to-charge ratio. As a step further to identification, the peptides can be sequenced with successive phases of fragmentation and mass analysis. This is a technique of tandem mass spectroscopy (MS/MS), in which a peptide has to pass through two analyzers for sequence determination. In the first analyzer, the peptide is fragmented by physical means generating fragments with nested size differing by only one amino acid. The molecular masses of these fragments are more precisely determined in the second analyzer yielding the sequence of fragments.

The first mass spectrometer was developed in 1899, but it took almost a century for the instrument to become a routine analytical method in proteomic research when fast atom bombardment ionization was developed, followed shortly by soft desorption/ionization methods, such as MALDI and electrospray ionization, to volatize biomolecules with masses of tens of kiloDaltons (kDa) into the gas phase under vacuum pressure without destroying them [13]. Thereafter, other soft ionization techniques that offered ambient conditions were also introduced, such as atmospheric pressure MALDI, direct analysis in real time, atmospheric-pressure solid analysis probe and hybrid ionization sources of MALDI and electrospray ionization (e.g., two-step fused droplet electrospray ionization, laser desorption atmospheric-pressure chemical ionization, electrosonic spray ionization, desorption electrospray ionization, and electrospray-assisted laser desorption/ionization). The five basic types of mass analyzers currently used in proteomic research are the quadrupole, ion trap, orbitrap, Fourier transform ion cyclotron resonance and TOF instruments, which differ in determining the mass-to-charge (m/c) ratios of the peptides. They have very different design and performance characteristics. These analyzers can stand alone or, in some cases, put together in tandem or in conjunction with ion mobility mass spectrometry to take advantage of the strengths of each. Several singly or multiply charged fragment ion types, such as b, y, a, c, z, v, y and immonium ions are produced in the gas phase of the spectrometer. In the bottom-up sequencing approach for protein identification in a shotgun proteomic experiment, proteolytic digestion of proteins is accomplished by cleavage of the different bonds along the peptide backbone and/or side chain through a charge-directed transfer to the vicinity of the cleavage side. These various mass spectrometers and the types of ions produced have become important analytical tools for studying and analyzing proteins, peptides and amino acids.

APPLICATION OF PROTEOMICS

One of the promising developments derived from the study of human genes and proteins has been the identification of the potential new drugs for the treatment of the disease. This relies on the genome and proteome information to identify protein associated with a disease. Computer softwares have used them as targets for new drug design. For example if a certain protein is implicated in a disease, its 3D structure provides the information to design drugs to interfere with the action of protein. A molecule that fits the active site of an enzyme will inactivate the enzyme. This is the basis of new drug discovery tools, which aims to find new drugs to inactivate proteins involved in disease. As genetic differences among individuals are found, researchers expect to use these techniques to develop personalized drugs that are more effective for the individual. A computer technique, which attempts to fit millions of small molecules to the three dimensional structure of a protein is called "virtual ligand screening". The computer rates the quality of the fit to various sites in the protein, with the goal of either enhancing or disabling the function of the protein, depending on the function of the cell.

In the early 1980s, researchers were not able to take advantage of structure-based methods in drug discovery process. This was due to number of factors the most important of which were a lack of computing power and docking programmes that could test potential models as well as lack of interest in the established community in part due to the aforementioned lack of tools. However, in 1980s, computing power and available programmes increased exponentially as well as the ability to obtain cheaper, more reliable X-ray crystallographic structures necessary for any type of computational study. This marked the start of new era and the first attempt and successes with two most famous being HIV-1 protease inhibitors and renin inhibitors (to combat hypertension). In contemporary drug discovery, structure based methods are an integral part of the drug development process. This change can be attributed to the rapid advances in genomics and structural biology, as well as developments in information technology.

The de novo approach involves constructing novel drug molecules from scratch based on the receptor/target structure. This is a particularly challenging task, considering that the search space of potentially feasible structure is to the order of 10^{100} [14]. The fundamentals of de-novo drug design include assembling drug like compounds, evaluating their potential and also searching the sample space for such compounds. The design of the drug molecule is generally driven by the structure of the receptor. Therefore, modeling the binding site is the greatest possible degree of accuracy is one of the critical steps in the process. Once the binding site is defined, the next step is to place the atoms or small molecule within it and evaluate the fit/interaction. This is done using a scoring function. Depending on the software package used, there now exists a wide variety of scoring functions ranging from simple steric constraints to function that approximate and evaluate binding free energy. The scoring functions are also responsible for guiding the growth and optimization of structures by assigning fitness values to the evaluated structures.

Different ways of constructing complete ligand molecule (insilico) include linking, growing random structure change and molecular dynamics based methods (Figure-3).

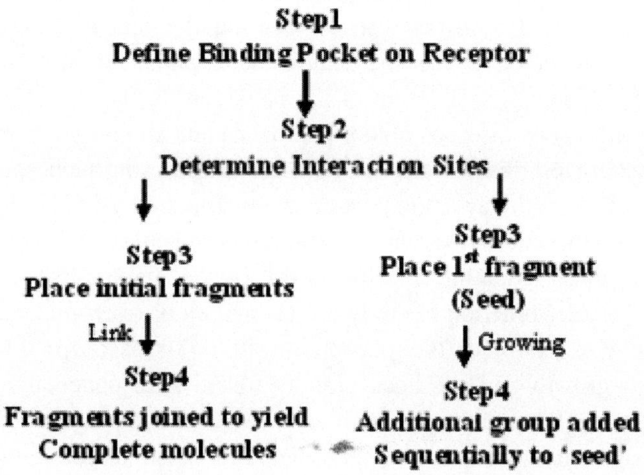

Figure 3. A flow chart showing different ways of constructing complete ligand molecules.

Linking

Atoms, functional groups or fragments are placed at pre determined interaction sites and are joined together to yield a complete molecule. The linking groups may be either pre-defined or generated on the fly with the constraints.

Growing

A single building block is chosen as a starting point or seed and placed in the interaction site of the receptor. Fragments or other groups are added to provide feasible interactions with the residues in the site. This process is continued until all fragments have been integrated into a single molecule.

Molecular Dynamics

The integral fragments are scattered randomly in the interaction site and then allowed to rearrange using molecular dynamics simulations. Scoring functions are used to evaluate the resultant structure. One of the major drawbacks of de novo algorithms is that they generally ignore synthetic feasibility of the designed structures.

Understanding the proteome, the structure and function of each protein and the complexities of protein–protein interactions will be critical for developing the most effective diagnostic techniques and disease treatment in the future. An interesting use of proteomics is using specific protein biomarkers to diagnose disease. A number of techniques allow testing for protein produced during a particular disease, which helps to diagnose the disease quickly. Techniques include western blot, immunohistochemical staining, enzyme linked immunosorbant assay (ELISA) or mass spectroscopy.

The ideal drug target for structure based drug design should bind a small molecule and should be closely associated with the disease. The small molecule that either changes the function of the drug target or in case of pathogenic organism inhibits the function of the target. This will ideally lead to cell death of the pathogen. In the latter case, the drug target should only be present in the disease cells or pathogens and should have unique function that allows for and encourages this selectivity. Furthermore, the uniqueness of the drug target guarantees that another pathway cannot restore the function of an inhibited target. The structure-based search for anti-cancer and autoimmune drugs is much more challenging, since the drug targets regulate essential cell functions. Hence, the targets are not unique and isolated, inhibition of their function not only affects mutant or over/ under activation cells but also normal cells. For example, the phosphoinositide 3-kinase (p13k) pathway is both involved in cellular growth and has been directly linked with pancreatic cancer activation. Trying to separate convulated pathways such as this makes the drug discovery process much more complex than with other diseases. Another option for targeting in cancer is DNA. Cisplatin and Bleomycin cross-link and cleave DNA, respectively, and are used to slow down cell division, particularly in cancerous tissues. The ligand-binding site of a drug target should be a pocket containing the hydrogen donors and acceptors, as well as hydrophobic residues. In many cases, the selected target location is the active site of an enzyme, as with sildenafil citrate (Viagra), which targets the catalytic subunit of NADPH (Jeremy); however, it can also be an assembly or regulatory site, as is the case with the phosphotransferase regulatory domain the *Bacillus subtilis* Spo Of protein, a histidine kinase (Dia-Fu). Even protein-protein interaction sites, which are often large and planar, have been selected as target sites (2-oxoglutarate) a naturally occurring molecule, affects the monomer-monomer interactions of GlnK, an ammonia transporter (Acr B) inhibitor.

A biomarker is a molecule, indicator, or test that can be used to measure such processes as disease progression, infection stage and drug efficacy, as well as other biological functions. Although the most common biomarkers in use today are small molecules and proteins, the growing fields of pharmacogenetics and pharmacogenomics are attempting to utilize genotypes, haplotypes and single nucleotide polymorphisms as biomarkers. A biomarker should be able to accurately discriminate between normal and an infected condition with high reproducibility. The challenge in proteomics research is to identify unique biomarkers from complex biological mixtures that unequivocally correlate with disease condition. Biomarkers can be utilized for many purposes. Also, established biomarkers are useful as risk factor indicators, capable of providing information to show that a person is susceptible to a disease. QT prolongation, a measure of change in the ventricular electrical cycle, is used as a assessment of a patients chances for survival after heart attack, as is tropionin T, a cardiac enzyme whose levels rise following heart attack. 5 hydroxytryptophan is a metabolic precursor to serotonin, and is found to localize around neoplasms in neuroendrocrine tissue. Labeling of these molecules allows for visualization of these events by fluorodeoxyglucose position emission tomography (FDG-PET), which is used in visualizing many types of malignancies. Another PET application uses SPA-RQ, a neurokinin-1 binder of the drug AprepitantTM, a drug used in the chemotherapy of patients to control vomiting [15].

There are currently many more biomarkers being studied and in development, and this number will continue to grow as our understanding of the body evolves. The combination of tracer molecules and imaging techniques, as in 5-hydroxytryptophan, creates a powerful method for visualizing areas that should not be touched surgically except as last resort. The growing use of proteomics to find biomarkers is important as well. Recent advances in analytical techniques such as mass spectrometry and column chromatography have allowed for more comprehensive studies of protein expression. In addition, 2 dimensional electrophoresis provides an over all view of expression, in similar fashion to gene studies. Proteomics studies using these techniques have identified ovarian cancer, macular degeneration, and lipoprotein composition. The following are some of the diseases that have characteristics biomarkers that physicians can use for diagnosis.

Alzheimer's Disease

In Alzheimer's diseases, elevation in beta secretase create amyloid/ beta- protein, which causes plaque to build up in the patients brain, which is thought to play a role in dementia. Targeting this enzyme decreases in the amyloid/ beta- protein and so slows the progression of the disease. A procedure to test for the increase in amyloid/beta protein is immunohistochemical staining in which antibodies bind to specific antigens or biological tissue of amlyloid/beta-protein.

Heart Disease

Heart disease is commonly assessed using several key protein based biomarkers. Standard protein biomarkers for CVD include interleukin-6, interleukin-8, serum amyloid A protein, fibrinogen and troponins. C TnI cardiac troponin I increases in concentration within 3 to 12 hours of initial cardiac injury and can be found elevated days after an acute myocardial infarction. A number of commercial antibody based assays as well as other methods are used in hospitals as primary tests for acute MI.

Cancer Therapy

Cancer, being a major healthcare concern worldwide, is one of the main targets for the application of emerging proteomic technologies and these tools promise to revolutionize the way cancer will be diagnosed and treated in the near future [16]. Today, as a result of the unprecedented advances that have taken place in molecular biology, cell biology and genomics there is a pressing need to accelerate the translation of basic discoveries into clinical applications. This need, compounded by mounting evidence that cellular model systems are unable to fully recapitulate all biological aspects of human disease, is driving scientists to increasingly use clinically relevant samples for biomarker and target discovery. Tissues are heterogeneous and as a result optimization of sample preparation is critical for generating accurate, representative, and highly reproducible quantitative data. Although a large number of protocols for preparation of tissue lysates has been published, so far no single recipe is able to provide a "one-size fits all" solubilization procedure that can be used to analyze the same lysate using different proteomics technologies. Here we present evidence showing that cell lysis buffer 1 (CLB1), a lysis solution commercialized by Zeptosens [a division of Bayer (Schweiz) AG], provides excellent sample solubilization and very high 2D PAGE protein resolution both when using carrier ampholytes and immobilized pH gradient strips. Moreover, this buffer can also be used for array-based proteomics (reverse-phase lysate arrays or direct antibody arrays), allowing the direct comparison of qualitative and quantitative data yielded by these technologies when applied to the same samples. The usefulness of the CLB1 solution for gel-based proteomics was further established by 2D PAGE analysis of a number of technically demanding specimens such as breast carcinoma core needle biopsies and problematic tissues such as brain cortex, cerebellum, skeletal muscle, kidney cortex and tongue. This solution when combined with a specific sample preparation technique - cryostat sectioning of frozen specimens - simplifies tissue sample preparation and solves most of the difficulties associated with the integration of data generated by different proteomic technologies.

Mass spectroscopy (MS) characterization of proteins is highly dependent on bioinformatics analysis. Once the peptide mass fingerprints or sequences are determined, bioinformatics programs can be used to search for a protein in a database of theoretically digested protein. The purpose of the database search is to find exact or nearly exact matches. However in reality, protease digestion is rarely perfect, often generating partially digested products as a result of missed cuts at expected cutting sites. Peptides resulting from MALD1-MS (matrix-assisted laser desorption ionization) are also charged which increases their mass slightly. To increase the discriminatory ability of the database search, the search engine must allow some leeway in matching molecular masses of peptides in the cases of missed cuts and changed modifications. The user is required to provide as much information as possible as input. For example, molecular masses of peptide fingerprints, peptide sequence, MW and pI of the intact protein, even the species names are important in obtaining unique identification of particular protein. A basic requirement for peptide identification through database matching is the availability of all the protein sequences from an organism. Thus, this method only works well with model organisms that have completely sequenced and well-annotated genomes, but has much limitation to be applied in nonmodel organisms.

Ex PASY (www.expasy.ch/tools/) is a comprehensive proteomics web server with a suite of programs for searching peptide information from the SWISS-PROT and TrEMBL databases. There are twelve database search tools in this server dedicated to protein identification based on MS data. For example, the AACompIdent program identifies protein based on pI. MW and amino acid composition and compares these values with theoretical composition of all proteins in SWISS-PROT/ TrEMBL. The number of candidate proteins can be further narrowed down by using species names and keywords. The TagIdent program can narrow down the candidate list by peptide sequences because of the high specificity of short sequence matches. The PeptIdent program incorporates mass fingerprinting information with information such as pI, MW and species name. Candidate proteins are ranked by the number of matching peptides. The Combsearch tool takes advantage of the strength of multiple parameters by using combined composition, sequence tags and fingerprinting information to perform combined searches against the database.

ProFound (http://prowl/.rockefeller.edu/profound.bin/WebProFound.exe) is a web server with a set of interconnected programs. It searches a protein sequence database using MS fingerprinting information. A Bayesian algorithm ranks the database matches according to the probability of database sequences producing the peptide mass fingerprints. Mascot (www.matrixscience.com/search-form-select.html) is another web server that identifies proteins based on peptides mass fingerprints, sequence entries, or raw MS/MS data from one or more peptides.

CONCLUSION

There's little doubt that proteomics, which is the study of an organism's complete complement of proteins, will have great impact in all areas of the life sciences in the years to come. In order to really understand biological processes, we need to understand how proteins function in and around cells since they are the basic functioning units. The task of studying the proteome has its share of challenges. One involves the sheer number of proteins that need to be identified. The 35,000 genes in the human genome can code for at least ten times as many proteins; in extreme cases a single gene alone can code for over 1,000. Another challenge is that amino acids, the building blocks of proteins, are so small. Each amino acid is made from 7 to 24 atoms. This is far beyond the reach of even the most powerful microscopes.

Proteomics technologies will play an important role in drug discovery, diagnostics and molecular medicine because it is the link between genes, proteins and disease. As researchers study defective proteins that cause particular diseases, their findings will help to develop new drugs that either alter the shape of a defective protein or mimic a missing one. Already, many of the best selling drugs today either act by targeting proteins or are proteins themselves. Advances in proteomics may help scientists to eventually create medications that are "personalized" for different individuals to be more effective and lesser side effects. Current research is looking at protein families linked to genes including cancer, diabetes and heart disease. Identifying unique patterns of protein expression, or biomarkers, associated with specific diseases is one of the most promising areas of clinical proteomics.

REFERENCES

[1] Anderson NL & Anderson NG (1998). Proteome and proteomics, new technologies new concepts and words. *Electrophoresis*, 19, 1853-1861.

[2] Blackstock WP & Weir MP (1999). Proteomics quantitative and physical mapping of cellular proteins. *Trends in Biotechnology,* 17, 121-127.

[3] James P (1997). Protein identification in the post-genome era: the rapid rise of proteomics. *Quarterly Reviews of Biophysics,* 30, 279-331.

[4] Marc R, Wilkins, Pasquali C, Appel RD, Ou K, Golaz O, Sanchez J-C, Yan JX, Gooley AA, Hughes G, Humphery-Smith I, Williams KL and Denis F. Hochstrasser DF (1996). From proteins to proteomes: Large scale protein identification by two dimensional electrophoresis and amino acid analysis. *Nature Biotechnology*, 14, 61-65.

[5] UNSW Staff Bio: Professor Marc Wilkins (http://www.babs.unsw.edu.au/directory. php?personnel ID= 12).

[6] Rogers S, Girolami M, Kolch W, Waters KM, Liu T, Thrall B & Willey HS. (2008). Investigating the correspondence between transcriptomic and proteomic expression profiles using coupled cluster models. *Bioinformatics*, 24, 2894-2900.

[7] Dhingraa V, Gupta M, Andacht T and Fu ZF 2005).New frontiers in proteomics research: A perspective. *International Journal of Pharmaceutics,* 299, 1-18.

[8] Steven B (2009).The major world of microRNAs. (http://www.nature.com/horizon/ rna/ background/micrornas.html).

[9] Olsen Jv, Blagoev B, Gnad F, Macek B, Kumar C, Mortensen P & Mann M (2006). Global *in vivo*, and site specific phosphorylation dynamics in signaling networks. *Cell*, 127, 635-648.

[10] Belle A, Tanay A, Bitincka L, Shamir R & O'Shea EK (2006). Quantification of protein half-lives in the budding yeast proteome. *Proceedings of National Academy of Sciences,* USA 103, 13004-13009.

[11] Serebriiskii I, Khazak V & Golemis EA [1999]. A two hybrid dual bait system to discriminate specificity of protein interactions. *Journal of Biological Chemistry,* 274, 17080-17087.

[12] Uetz P, Giot L, Cagney G, Mansfield TA, Judson RS, Knight JR, Lockshon D, Narayan V, Srinivasan M, Pochart P, Qureshi-Emili A, Li Y, Godwin B, Conover D, Kalbfleisch T, Vijayadamodar G, Yang M, Johnston M, Fields S, Rothberg JM &.Rothberg JM (2000). A comprehensive analysis of protein-protein interactions in *Saccharomyces cerevisiae* . *Nature*, 403, 623-627.

[13] Ahmad FE (2009). Utility of mass spectrometry for proteome analysis: part II. Ion-activation methods, statistics, bioinformatics and annotation. *Expert Review of Proteomics*, 6:2, 171-197.

[14] Bohacek RS, C McM & Guida WC (1996). The art and practice of structure based drug design: a molecular modeling prespective. *Medical Research Review,* 16, 3-50.

[15] Frank R & Hargreaves R (2003). Clinical biomarkers in drug discovery. *Nature Reviews on Drug Discovery,* 2, 566-580.

[16] Gromov P, Cilis JE, Gromova I, Rank F, Timmermans-Wielenga V & Moreira JM (2008). A single lysis solution of tissue samples by different proteomic technologies.Mol. *Oncology*, 2, 368-379.

EXTERNAL LINKS

Proteomics (http://www.dmoz.org//science Biology/ Biochemistry- and –Molecular Biology/ Biomolecules/ Pro at Open Dictionary Project).

Proteomic depletion, antityprisin: (http://www.youtube.com/watch? v= Oc Zpp1gYq-g) PAMP The Proteolysis Map-animation.

PatternLab for proteomics: a tool for analyzing shotgun proteomic spectral counting data (http://pcarvalho.com/patternlab).

Protein Information Crawler (PIC) (http://www.zoo.uni-heidelberg.de/mfa/PIC/index.htm)- Software for extensive spidering of multiple protein information resources for large protein sets.

Springer Protocols in proteomics (http://www.springerprotocols.com/).

In: Recent Trends in Biotechnology, Volume 1
Editors: M.P. Singh, A. Agrawal and B. Sharma
ISBN 978-1-60876-148-7

Chapter 3

ENRICHING BIOTECHNOLOGY WITH RECENT BIOINFORMATICS TOOLS AND PROTOCOLS

Shailendra K. Gupta*[1], *Suchi Smita and Qamar Rahman
[1]Indian Institute of Toxicology Research, Lucknow (UP) India
Integral University, Lucknow (UP) India

ABSTRACT

The development of databases and the use of bioinformatics have not only made science interesting but also easier to carry out new researches using the existing biological data. Enormous amount of the available data could be sorted out with ease as per the demand of the topic of the research using various algorithms, machine learning approaches and information retrieval systems. Genomics and Proteomics are an area of research which have still many mysteries unsolved and here the approaches of bioinformatics has made the tedious work much more simpler by providing numerous popular tools like BLAST and FASTA, ClustalW, PSI-Blast etc. to sort out the aggregate features from the completely sequenced genomic data. New methodologies are being developed to identify distant homologs, presence and absence of a protein in any sequenced genome, prediction of gene functions and many more. The role of bioinformatics in determining the complex structure of proteins, *de novo* modeling of loops in protein structure is worth mentioning. The revolution of bioinformatics in drug designing has given the hope to conquer the deadly diseases. The development of novel enzyme inhibitors like HIV-1 protease has now become possible. A number of programs for prediction of various ADME (Adsorption, distribution, metabolism and excretion) properties of hit molecules are successfully designed. Various computer-assisted drug discovery programs were developed for a wide range of applications including rigid body protein ligand docking, flexible docking and protein-protein interactions.

INTRODUCTION

After the first algorithm published by Needleman and Wunsch in 1970 for comparison of nucleotide and protein sequences, Bioinformatics became the centre of modern biological research. Availability of large number of completely sequenced prokaryotic and eukaryotic genomes changed the way we hypothesized and practiced biology in the past decades. The rapid instrumental growth in High throughput techniques for genome sequencing, gene expression, and protein profiling stimulates the development of new algorithms and computational tools for acquiring, storing, analyzing and representing biological information. It is the amount and type of biological data, about the metabolic networks and molecular interactions that have driven the rapid advancement of bioinformatics.

Database Resources For Storing Biological Knowledge

With the first biological database developed in 1960s, when M. Dayhoff and colleagues at Protein Information Resource (PIR) collected all the protein sequences at that time and published the work as *Atlas of Protein Sequence and Structure* [1], biological databases became the centre for knowledge sharing and interpretation. Since 1993, the journal of Nucleic Acid Research (NAR) publishes one special database issue on all available biological databases every year in the month of January. These issues are freely accessible from the internet (http://nar.oxfordjournals.org/archive) [2] and provide important information to work with biological databases. In the 2009 Nucleic Acid Research annual database issue, not only the total number of databases in the collection increased to 1170 but the list of countries represented the online collection has also increased through the inclusion of the first Argentinean database TcSNP for genetic variation in *Trypanosoma cruzi* (http://snps.tcruzi.org) [3]. Depending on the nature and scope of information included, different categories and sub categories of biological databases can be distinguished (Table 1). Biological databases can be roughly divided into three important categories: Primary, Secondary and Specialized database. While the primary databases contain original biological data generated after experimental results along with some interpretations, secondary databases store computationally processed or manually curated information. Specialized databases are being intended to focus as a particular research.

Knowledge Generation Using Recent Data Mining Techniques

To extract and interpret useful biological information and pattern from noisy data produced by high-throughput technologies is a demanding task for bioinformatics. Among many techniques emerged under the same goal of mining knowledge form the data in past 10 years, various algorithms, databases searching, machine learning approaches and information retrieval systems are essential and successfully applied to attempt many problems in Bioinformatics. BioMart (http://www.biomart.org) is one of the important query-oriented data mining systems developed jointly by the Ontario Institute for Cancer Research (OICR) and the European Bioinformatics Institute (EBI). BioMart is an open source data management system that comes with a range of query interfaces that allow users to group and refine data based upon many different criteria [4]. BioMart is popular because of its easy integration with many external software packages such as the Distributed Annotation System (DAS) [5], Galaxy [6], Cytoscape [7] and Taverna [8].

Table 1. Categories and subcategories of databases are listed in column 1 followed by total no. of databases with few examples as appeared in the 2009 Nucleic Acid Research annual database issue

Categories / sub categories of Databases	No. of Databases	Examples
Nucleotide Sequence Databases		
International Nucleotide Sequence Database Collaboration	3	DDBJ, EMBL, GenBank
Coding and non-coding DNA	43	RefSeq, Islander, FREP, UniVec
Gene structure, introns and exons, splice sites	24	ASDB, SpliceDB, Yeast Intron Database
Transcriptional regulator sites and transcription factors	64	cisRED, DBTSS, TRANSFAC
RNA sequence databases	72	FlyRNAi, MiRNAMap, RNAdb
Protein Sequence Databases		
General sequence databases	14	PIR, PRF, Swiss Prot, UniParc
Protein properties	16	BindingDB, Cybase, PPD
Protein localization and targeting	23	LOCATE, NPD, PeroxisomeDB
Protein sequence motifs and active sites	25	PDBSite, PRINTS, PROSITE, Blocks
Protein domain databases; protein classification	38	COG, InterPro, Pfam, ProDom
Databases of individual protein families	73	ABCdb, Animal toxin Database, Knottin Database
Structure Databases		
Small molecules	18	DrugBank, Ligand, PubChem
Carbohydrates	9	GlycoSuit, CCSD, GlycoSuitDB
Nucleic acid structure	15	Greglist, RNA FRABASE, SCOR
Protein structure	84	CATH, Dali, MMDB, PDB
Genomic Databases		
Genome annotation terms, ontologies and nomenclature	12	BioThesaurus, GO, TreeBase
Taxonomy and identification	11	GeneTrees, MICdb, RISSC
General genomics databases	45	BacMAP, EBI Genomes, Entrez Genomes, KEGG
Viral genome databases	28	HCVDB, HIV Sequence Database, VirGen
Prokaryotic genome databases	68	EchoBase, EcoCys, GenomeAtlas, Strepto DB
Unicellular eukaryotes genome databases	19	GeneDB, ProtozoaDB, PlasmoDB
Fungal genome databases	31	CandidaDB, GermOnline, SGD
Invertebrate genome databases	54	FlyBase, VectorBase, WormBase

Table 1. (Continued)

Categories / sub categories of Databases	No. of Databases	Examples
Metabolic and Signaling Pathways		
Enzymes and enzyme nomenclature	13	BRENDA, ENZYME, ExplorEnz
Metabolic pathways	23	BioCarta, BioCyc, LIGAND, MetaCyc
Protein-protein interactions	77	BioGRID, MINT, Reactome, SMART
Signaling pathways	6	NetworKIN, PID, PRRDB
Human and Other Vertebrate Genomes		
Model organisms, comparative genomics	68	Ensembl, HomoloGene, KOG, MGD, RatMap
Human genome databases, maps and viewers	16	GeneCards, GeneLoc, UCSC Genome Browser
Human ORFs	28	GeneAnnot, ORFDB, UniGene
Human Genes and Diseases		
General human genetics databases	15	GenAtlas, OMIM, SNP2NMD
General polymorphism databases	32	dbSNP, HapMap, PharmGKB, SNPlogic
Cancer gene databases	25	CancerGenes, OncoDB.HCC, TGDBs
Gene-, system- or disease-specific databases	56	HemBase, MDPD, SynDB
Microarray Data and other Gene Expression Databases	67	GeneNote, GENSAT, M3D, OncoMine
Proteomics Resources	20	2D-PAGE, PeptideAtlas, PRIDE
Other Molecular Biology databases	10	BioModels, CycleBase, PubMed
Drug and drug design	22	DrugBank, GLIDA, SuperDrug, TTD
Molecular probes and primers	10	AOBase, MPDB, Primer Studio, VirOligo
Organelle Databases	8	Organelle DB, Plant Organelle Database
Mitochondrial genes and proteins	18	HMPD, Mitome, MitoRes, YMPD
Plant Databases		
General plant databases	42	AgBase, GeneFarm, PlantGDB, PPDB
Arabidopsis thaliana	27	AtGDB, AtPID, MAtDB, DATF
Rice	18	Oryzabase, Rice Proteome Database
Other plants	18	ForestTreeDB, MaizeGDB, SoyBase
Immunological databases	27	AntiJen, IMGT, MHCBN, MHCPEP

Data mining can be divided into two broad groups; supervised and unsupervised techniques. While supervised algorithms require a posteriori knowledge and experience with the data, unsupervised algorithms are based on a priori techniques and do not need knowledge. Usually such statistical approaches lag the ability to detect non-linear relations but provide understandable results. However, new advances in artificial neural networks and genetic algorithms support the pattern recognition process to find non-linear relations. Classification and decision trees are the example of a posteriori knowledge and clustering is the techniques used to detect similarity, a priori [9]. Many applications in bioinformatics are based on classification and clustering method viz. functional annotations, protein homology modeling, protein family classification and metabolic pathway clustering.

Advancement In Functional Annotation Techniques

Over the past 30 years, many computational algorithms centered around one of the important problem i.e. annotation of new nucleotide and protein sequences. Among them, dynamic programming algorithms by Smith and Waterman (1981) [10] and its heuristic variations BLAST [11] and FASTA [12] were based on the sequence similarity. These algorithms were based on the fact that biological sequences are conserved during evolution. Novel functionalities in the organisms do not appear because a new gene has suddenly arisen but are due to continuous development and modification during evolution. Based on these facts, functional annotation can be transferred to new proteins if there is significant sequence similarity.

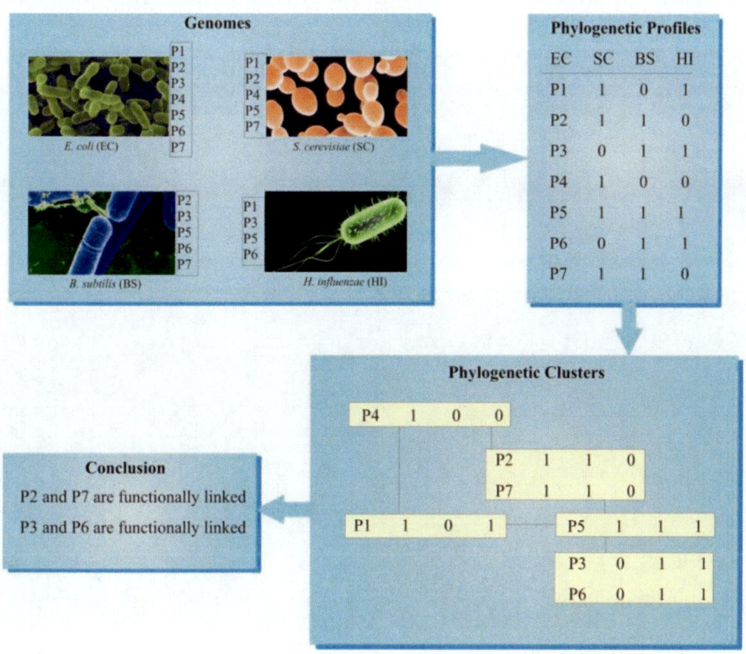

Figure 1. Schematic representation of protein phylogenetic profiles for the hypothetical case of four fully sequenced genomes (from *E. coli*, *Saccharomyces cerevisiae*, *Haemophilus influenzae*, and *Bacillus subtilis*) focused on seven proteins (P1-P7). For each E. coli protein, profile was constructed, indicating which genome codes for homologous protein. Proteins that share same profile were clustered. Proteins with identical (or similar) profiles are boxed to indicate that they are likely to be functionally linked. Clusters connected by lines have phylogenetic profiles that differ by one bit and are termed *neighbors*.

Other sophisticated methods for more accurate and reliable functional annotation use multiple sequences from protein families with similar functions. Profiles [13], hidden markov models [14], PSI-BLAST [15], Profile-Profile [16] are few popular methods to cater the aggregate features from whole sequences for protein families. These aggregate features are represented as pattern or motif. Methods based on patterns and motifs include BLOCKs [17], MEME [18], GIMSAN [19] and eMotif [20].

Recent methodologies for predicting protein function are based on DNA microarray gene expression data, phylogenetic profiles and genetic networks. These methods are widely used to detect distant homologies even when the sequence similarity is less than 30%. Using phylogenetic profile methods, presence or absence of protein can be sensed in every sequenced genome for inferring the functional relationship among proteins as shown in Figure-1 [21]. Phylogenetic profile and genetic network methods have been widely used in gene function prediction, gene-phenotype prediction, protein co-evolution analysis and prediction of protein functional annotation.

PROTEIN STRUCTURE PREDICTION TECHNIQUES

Because of the complexity in structure determination of proteins, the gap between known protein sequences and structures is increasing day by day (Figure-2). The structure prediction problem is of great importance as it is a prerequisite for many biological analysis.

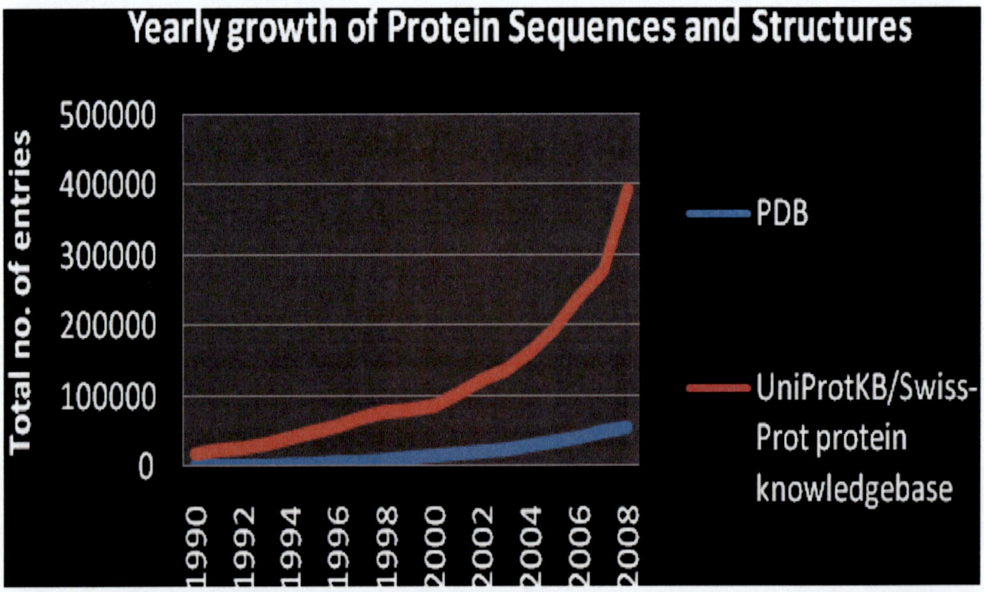

Figure 2. Yearly growth of UniProtKB/ Swiss-Prot Protein knowledgebase and Protein Data Bank. Increasing gaps between two clearly indicate the advancement in high throughput protein sequencing techniques in comparison to structural determination of proteins.

Table 2. List of software / webserver available for protein 3D structure prediction grouped into 3 categories on the basis of their working methodologies i.e. Comparative Modeling (Homology modeling), Threading / Fold recognition and *Ab initio* Prediction approach

Sever/ Tool	Method	Description	URL	Organisation
Comparative modeling based tool				
3D-JIGSAW	Fragment assembly	Automated webserver	http://bmm.cancerresearchuk.org/~3djigsaw/	Cancer Research UK London Research Institute
ESyPred3D	Template detection, alignment using neural network, 3D modeling	Automated webserver	http://www.fundp.ac.be/sciences/biologie/urbm/bioinfo/esypred/	University of Namur
Geno3D	Satisfaction of spatial restraints	Automated webserver	http://geno3d-pbil.ibcp.fr/cgi-bin/geno3d_automat.pl?page=/GENO3D/geno3d_home.html	Institut de Biologie et Chimie des Protéines, France
HHpred	Template detection, alignment, 3D modeling	Webserver	http://toolkit.tuebingen.mpg.de/hhpred	Max-Planck Institute for Developmental Biology
MODELLER	Satisfaction of spatial restraints	Standalone program in Python	http://salilab.org/modeller/	University of California
ROBETTA	Rosetta homology modeling and ab initio fragment assembly with Ginzu domain prediction	Webserver	http://robetta.org/	University of Washington
SWISS-MODEL	Local similarity/fragment assembly	Automated webserver	http://swissmodel.expasy.org/	Swiss Institute of Bioinformatics
Methods based on Threading / fold recognition				
3D-PSSM	3D-1D sequence profiling	Webserver	http://www.sbg.bio.ic.ac.uk/~3dpssm/	Biomolecular Modelling Laboratory, Imperial Cancer Research Fund, England

Table 2. (Continued)

Sever/ Tool	Method	Description	URL	Organisation
Methods based on Threading / fold recognition				
I-TASSER	Combination of several methods	Webserver	http://zhang.bioinformatics.ku.edu/ I-TASSER/	University of Kansas
SUPERFAMILY	Hidden Markov modeling	Webserver/ standalone	http://supfam.org/SUPERFAMILY	Laboratory of Molecular Biology, MRC, Cambridge
GenTHREADER	sequence profile based fold recognition method	Webserver	http://bioinf.cs.ucl.ac.uk/psipred/	Department of Computer Science University College London
Ab initio prediction methods				
CABS	Reduced modeling tool	Downloadable program	http://www.biocomp.chem.uw.edu.pl/services.php	Dept. of chemistry, University of Warsaw, Poland
Bhageerath	Modeling and predicting protein structures at the atomic level	Webserver	http://www.scfbio-iitd.res.in/bhageerath/index.jsp	Dept. of information technology, IIT, Delhi, India

To evaluate the progress of all variants of computational protein structure prediction methods, a large scale community based experiments CASP (Critical Assessment of Protein Structure Prediction) is conducted every two years, since 1994. There has been substantial progress in comparative modeling approaches, where the model of protein sequence is determined using the information from an evolutionarily related structural template. Template free modeling, although still very approximate, is now producing topologically correct models for some small proteins [22]. Some of the servers and tools for protein structure prediction are listed in Table 2. Among them, modelling is very popular not only for comparative protein structure modeling by satisfaction of spatial restraints [23, 24] but also for de novo modelling of loops in protein structures, optimization of various model of protein structures, multiple alignment of protein sequences and structures, clustering etc. One example of modeling is shown in Figure 3 using the data files available at http://salilab.org/modeller , where the structure of lactate dehydrogenase from *Trichomonas vaginalis* is modelled using the suitable template structure available in protein data bank, bacterial malate dehydrogenase (PDB code 1bdmA) in this case. The loop region is not properly modeled, most probably because of its participation in the active site formation and in the absence of ligand, inreliable coordinates in the pdb file. The loop modeling may be carried out by using multiple templates, *ab-initio* methods or by selecting template where lignad is bound to binding sites.

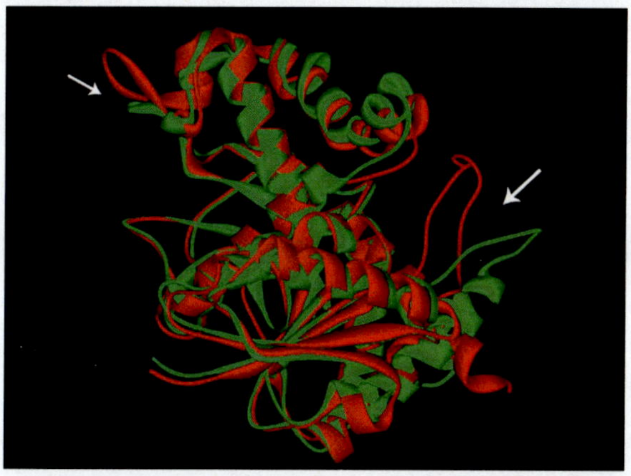

Figure 3. An example of a comparative modeling approach to model lactate dehydrogenase enzyme from *Trichomonas vaginalis* based on bacterial malate dehydrogenase (PDB code 1bmdA). The best model is Red, and the available template structure (45% sequence identity to the target, PDB code 1bdmA) is green. Target and template are superimposed to show the accuracy of modeled structure. Helixes and sheets are properly modeled. Arrow indicates the loop region, poorly modeled, because of unreliability in the template coordinates.

Strategies For Computer-Assisted Drug Discovery (CDD)

Bioinformatics completely revolutionized the traditional process of drug designing. Virtual screening followed by in vitro experiment is recognized as a promising approach in today's drug discovery projects. Computer assisted drug designing processes successfully

Table 3. List of some prominent computer-assisted drug discovery programs

Program	Description	URL
Protein – Protein docking		
3D Dock Suite	FTDock, RPScore and MultiDock Perform rigid body docking of 2 biomolecules	http://www.bmm.icnet.uk/docking
ClusPro	Fully automated protein-protein docking. Perform rigid docking by DOT, ZDOCK, and PIPER. Clustering of complexes	http://structure.bu.edu/Projects/PPDocking/cluspro.html
DOT	Rigid docking of macromolecules including Protein, DNA, RNA	http://www.sdsc.edu/CCMS/DOT/
FireDock	Refinement and re-scoring of rigid-body protein-protein docking solutions	http://bioinfo3d.cs.tau.ac.il/FireDock/
HEX	Protein docking and molecular superimposition program	http://www.loria.fr/~ritchied/hex/
HADDOCK	High Ambiguity Driven biomolecular DOCKing	http://www.nmr.chem.uu.nl/haddock/
ZDock	Fast Fourier Transform based protein docking program	http://zdock.bu.edu/
Protein – ligand docking		
Autodock	Rigid & flexible docking. Docking of the ligand to a set of grids describing the target protein	http://autodock.scripps.edu/
DOCK	Dock small molecules or fragments, allow flexible ligand docking, shape based algorithm	http://dock.compbio.ucsf.edu/
FlexX	Incremental construction approach of ligand and allow flexible protein-ligand docking	http://www.biosolveit.de/FlexX/
GLIDE	Complete suite for protein-ligand interaction based on systematic search techniques	http://www.schrodinger.com/
GOLD	Uses genetic algorithms for docking	http://www.ccdc.cam.ac.uk/products/life_sciences/gold/
LigPlot	For automatically plotting protein-ligand interaction	http://www.biochem.ucl.ac.uk/bsm/ligplot/ligplot.html
VEGA	Rank ligand on the basis of receptor-ligand interaction energy	http://users.unimi.it/~ddl/
Discovery Studio	Flexible docking, virtual high throughput screening, *de novo* ligand generation	http://www.accelrys.com

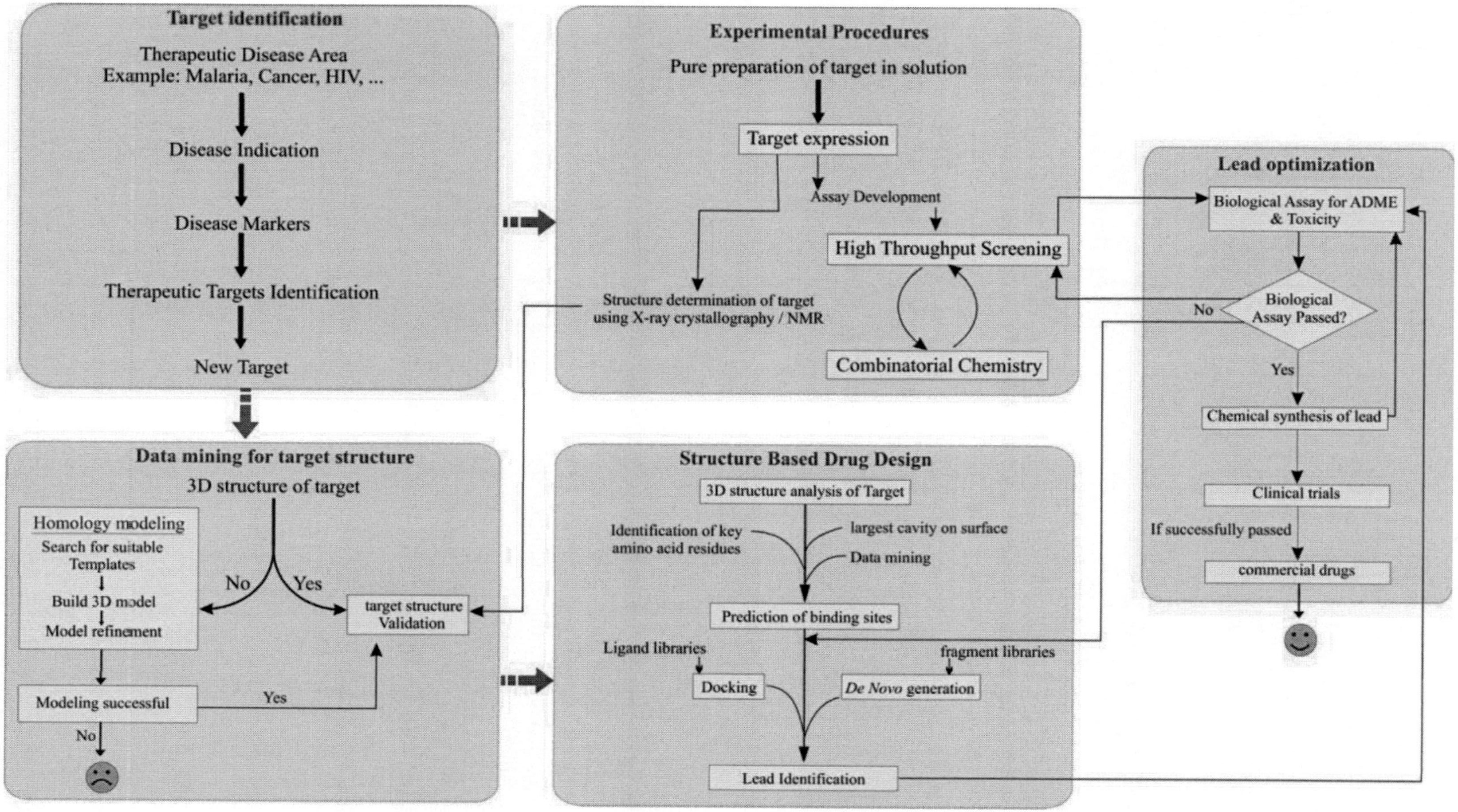

Figure 4. Flow chart for computer-assisted drug designing process

contributed to develop numerous novel enzyme inhibitors, including inhibitors of thymidylate synthase [25], HIV-1 protease [26], Dihydrofolate reductase [27], Acetylcholinesterase [28]. Several software and databases developed for computer assisted drug discovery are illustrated in Table 3 and the process is summarized in Figure 4. The process starts with the target identification and its structural analysis. The ideal target is one that is closely linked to human disease and binds with small molecules (ligand) in order to carry out a function. The field of genomics and proteomics contributed greatly to identify novel potential targets retrieved by comparing whole genome of host and pathogen over the last several years.

Analysis of 3D Structure & Identification of Target Sites

The process of CDD initiates with the detailed structural analysis of target molecule and prediction of potential ligand binding sites on its surface. The problem to detect putative binding sites can be approached in several ways. The common approach is to determine the cavities in the protein structure and assume the biggest cavity as the active site. More precise method searches for key amino acid residues involved in catalysis or cofactor, and are considered to form the binding sites. However, in the absence of experimental data, uncertainty remains in the accurate prediction of binding sites. Target sites for protein-protein interaction are difficult to identify as they are often flat, large and hydrophobic [29].

Drug Designing Methodology

Once the binding site is determined, thousands of compounds can be tested using high-throughput screening technologies, although the better way is to select lead compounds using computer driven assays. Molecular databases can be virtually screened to identify small molecules that are docked in the target sites, out of millions of molecules available. De novo generation approach is initiated by placing small fragments of molecules, such as benzene ring, amino group etc. into the binding cavity. These small fragments further linked among themselves to form best inhibitors.

Modification and Evaluation of Lead Compounds

Methods for modification of lead compounds are meant to enhance activity, detect toxicity, increase pharmacokinetics and reduce side effects. These methods are normally based on structural data mining to detect and analyze biological assays of similar compounds. Several programs are designed for prediction of various ADME properties which has to be tested later during preclinical evaluation for determination of suitability of hits into therapeutic agents [30].

In recent years, number of docking systems developed to optimize the prediction of binding affinity using molecular mechanics force field methodologies. One most commonly used system is Autodock [31] developed by G. M. Morris and co-workers. The system uses grid based techniques in which the interaction energy between the atoms of receptor & ligand are pre-calculated as the points of a grid. Hybrid genetic algorithm is used to adjust the position and confirmation of ligand in the binding cavity of protein. DOCK [32] and Discovery Studio [33] are based on a negative image of binding site, in which a sphere is generated around the cavity. During the process, virtual libraries are scanned to find the ligand that matches to the sphere. For all the ligands that fit in the sphere, full estimation of binding affinity is computed and ligands are arranged on the basis of different scoring functions. FlexX [34] uses a fragmentation approach in which ligand is fragmented and incrementally reconstructed in iterative manner using the tree-search algorithm in the binding site of receptor. GOLD [35] uses a template based on H-bond donors & acceptors of protein and uses genetic algorithm to sample over all possible combination of inter molecular hydrogen bond and ligand confirmations. One example of protein-ligand interaction is illustrated in Figure 5 using Autodock software, where the docking was performed between HIV-1 protease and nonpeptide cyclic ureas.

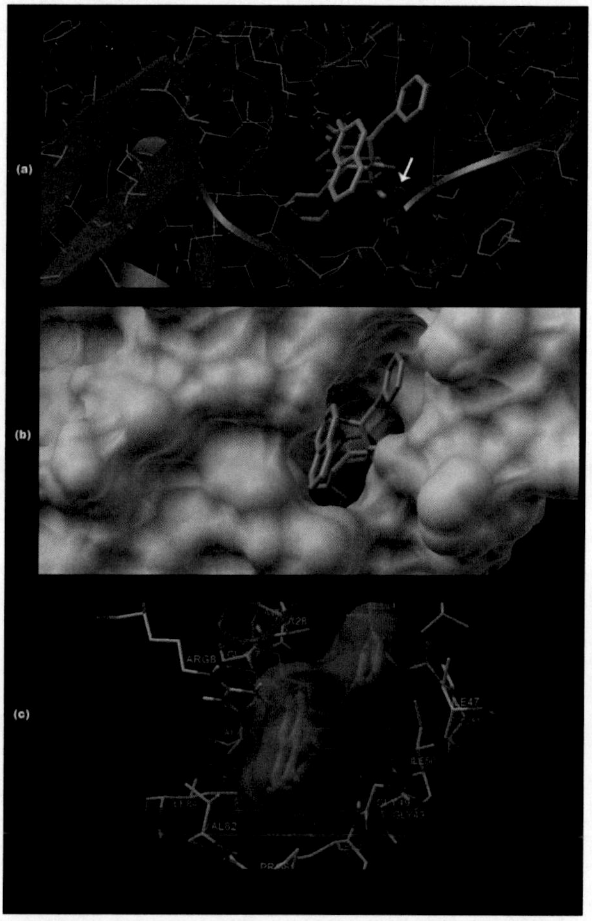

Figure 5. Nonpeptide cyclic ureas, HIV-1 protease inhibitors, in the binding cavity of protein (PDB id: 1hvr). (a) Protein is displayed in its secondary structure. Arrow shows the hydrogen bonding between protein-ligand. (b) A molecular surface of the protein is generated to properly illustrate the binding cavity. (c) Residues near binding site are displayed as wireframe to analyze their interaction with ligand molecule.

As structural genomics, proteomics, bioinformatics, computational power, knowledge of disease cycles and therapeutic targets continue to explode with new advances, further understanding in the structure-based drug designing are likely to follow.

Bioinformatics and System Biology

The quantity and range of omics data (Genomics, Proteomics, Transcriptomics, Metabolomics) along with recent technologies developed till date has made possible the research in Bioinformatics to move beyond the study of individual biological components (genes, proteins etc). The requirement is now to develop cooperation between individual datasets to design more realistic models for system biology in order to understand the functioning of the biological systems. Computer models performing complete cell simulation are more in demand [36]. Various computational environments were designed for studying simulation using System Biology Markup Language (SBML) like SimBiology Toolbox of MATLAB [37], SYCAMORE [38] and PROMOT [39].

Conclusion

Despite of some drawbacks, there is no doubt that bioinformatics is a field that has enormous capacity for revolutionizing biological research in the days to come. Along with providing more consistent and more precise computational tools for all the domain of biotechnology the biggest challenge for bioinformatics is to develop overall biology from a qualitative to a quantitative and predictive science. Many new upcoming areas where bioinformatics is going to play an important role are simulation of entire biological system, development of personalized drugs, identification of new strains for bioremediation and biofuel generation etc.

References

[1] Dayhoff, M. O. (1978). *Atlas of Protein Sequence and Structure.*Vol. 5, Suppl. 3. Washington, DC: National Biomedical Research Foundation.

[2] Galperin, M. Y. & Cochrane, G. R. (2009). Nucleic Acid Research annual Database Issue and the NAR online Molecular Biology Database Collection in 2009. *Nucleic Acid Research, 37,* D1-D4.

[3] Ackermann, A. A.; Carmona, S. J. & Aguero, F. (2009). TcSNP: a database of genetic variation in *Trypanosoma cruzi. Nucelic Acid Res., 37,* D544-D549.

[4] Smedley, D.; Haider, S.; Ballester, B.; Holland, R.; London, D.; Thorisson, G. & Kasprzyk, A. (2009). Biomart - biological queries made easy. *BMC Genomics, 10.*

[5] Dowell, R. D.; Jokerst, R.M.; Day, A.; Eddy, S. R. & Stein, L. (2001). The distributed annotation system. *BMC Bioinformatics, 2,* 7.

[6] Giardine, B.; Riemer, C.; Hardison, R. C.; Burhans, R.; Elnitski, L.; Shah, P.; Zhang, Y.; Blankenberg, D.; Albert, I.; Taylor, J.; Miller, W.; Kent, W. J. & Nekrutenko, A.

(2005). Galaxy: A platform for interactive large-scale genome analysis. *Genome Research, 15,* 1451-1455.

[7] Cline, M. S.; Smoot, M.; Cerami, E.; Kuchinsky, A.; Landys, N.; Workman, C.; Christmas, R.; Avila-Campilo, I.; Creech, M.; Gross, B.; Hanspers, K.; Isserlin, R.; Kelley, R.; Killcoyne, S.; Lotia, S.; Maere, S.; Morris, J.; Ono, K.; Pavlovic, V.; Pico, A. R.; Vailaya, A.; Wang, P. L.; Adler, A.; Conklin, B. R.; Hood, L.; Kuiper, M.; Sander, C.; Schmulevich, I.; Schwikowski, B. & Warner, G. J. (2007). Integration of biological networks and gene expression data using Cytoscape. *Nature Protocols, 2,* 2366-2382.

[8] Hull, D.; Wolstencroft, K.; Stevens, R.; Goble, C.; Pocock, M. R.; Li, P. & Oinn, T. (2006). Taverna: A tool for building and running workflows of services. Nucleic Acids Res, 34:W729-732.

[9] Hand D.; Mannila H. & Smyth P. (2001). *Principles of Data Mining.* Cambridge, MIT Press.

[10] Smith, T. F. & Waterman, M. S. (1981). Identification of common molecular subsequences. *Journal of Molecular Biology, 147,* 195-197.

[11] Altschul, S. F.; Gish, W.; Miller, E. & Lipman, D. J. (1990). Basic local alignment search tool. *Journal of Molecular Biology, 215,* 403-410.

[12] Pearson, W. (1990). Rapid and sensitive sequence comparison with FASTP and FASTA. *Meth. Enzymol., 183,* 63-98.

[13] Gribskov, M.; McLaclan, A. D. & Eisenberg, D. (1987). Profile analysis: Detection of distantly related proteins. *Proc. Natl. Acad. Sci. USA, 84,* 4355-4358.

[14] Krogh, A.; Brown, M.; Mian, I. S.; Sjolander, K. & Haussler, D. (1994). Hidden Markov models in computational biology: Applications to protein modeling. *Journal of Molecular Biology, 235,* 1501-1531.

[15] Altschul, S. F.; Madden, T. L.; Schaffer, A. A.; Zhang, J.; Zhang, Z.; Miller, W. & Lipman, D. J. (1997). Gapped BLAST and PSI-BLAST: A new generation of protein database search programs. *Nucleic Acid Res., 25,* 3389-3402.

[16] Sadreyev, R. & Grishin, N. (2003). Compass: A tool for comparison of multiple protein alignment with assessment of statistical significance. *Journal of Molecular Biology, 326,* 317-336.

[17] Henikoff, S. & Henikoff, J. G. (1994). Protein family classification based on search a database of blocks. *Genomics, 19,* 97-107.

[18] Grundy, W. N.; Bailey, T. L.; Elkan, C. P. & Baker, M. E. (1997). Meta-MEME: Motif –based hidden Markov Models of biological sequences. *Computer Applications in the Biosciences, 13,* 397-406.

[19] Patrick, N.; Uri, K. (2008). GIMSAN: A Gibbs motif finder with significance analysis. *Bioinformatics, 24 (19)*, 2256-2257.

[20] Nevill-Manning, C. G.; Wu, T. D. & Brutlag, D. L. (1998). Highly specific protein sequence motifs for genome analysis. *Proc. Natl. Acad. Sci USA, 95(11),* 5865-5871.

[21] Pellegrini, M.; Marcotte, E. M.; Thompson, M. J.; Eisenberg, D. & Yeates, T. O. (1999). Assigning protein functions by comparative genome analysis: Protein phylogenetic profiles. *Proc. Natl. Acad. Sci. USA, 96,* 4285-4288.

[22] Moult, J. (2005). A decade of CASP: progress, bottlenecks and prognosis in protein structure prediction. *Current Opinion in Structural Biology, 15(3),* 285-289.

[23] Marti-Renom, M. A.; Stuart, A.; Fiser, A.; Sanchez, R.; Melo, F. & Sali, A. (2000). Comparative protein structure modeling of genes and genomes. *Ann. Rev. Biophys. Biomolec. Struct., 29,* 291–325.

[24] Fiser, A.; Do, R. K. G. & Sali, A. (2000). Modeling of loops in protein structures. *Protein Science, 9,* 1753–1773.

[25] Tondi, D.; Slomczynska, U.; Costi, M. P.; Watterson, D. M.; Ghelli, S. & Shoichet, B. K. (1999). Structure-based discovery and in-parallel optimization of novelcompetitive inhibitors of thymidylate synthase. *Chemistry & Biology, 6(5),* 319-331.

[26] Sherman, W.; Tidor, B. (2008). Novel Method for Probing the Specificity Binding Profile of Ligands: Applications to HIV Protease. *Chemical Biology & Drug Design, 71(5),* 387-407.

[27] Wyss, P. C.; Gerber, P.; Hartman, P. G.; Hubschwerlen, C.; Locher, H.; Marty, H. & Stahl, M. (2003). Novel Dihydrofolate Reductase Inhibitors. Structure-Based versus Diversity-Based Library Design and High-Throughput Synthesis and Screening. J. Med. Chem., *46* (12), 2304–2312.

[28] Racchi, M.; Mazzucchelli, M.; Porrello, E.; Lanni, C. & Govoni, S. (2004). Acetylcholinesterase inhibitors: novel activities of old molecules. *Pharmacological Research, 50(4),* 441-451.

[29] Gadek, T. & Nicholas, J. (2003). Small molecule antagonists of proteins. *Biochem. Pharmacol, 65,* 1-8.

[30] Clark, D. E. & Pickett, S. D. (2000). Computational methods for the prediction of "drug likeness". *Drug Discovery Today, 5,* 49-58.

[31] Rosenfeld, R. J.; Goodsell, D. S.; Musah, R. A.; Morris, G. M.; Goodin, D. B. & Olson, A. J. (2003). Automated docking of ligands to an artificial active site: augmenting crystallographic analysis with computer modeling. *Journal of computer-aided molecular design, 17,* 525-536.

[32] Moustakas, D. T.; Lang, P. T.; Pegg, S.; Pettersen, E.; Kuntz, I. D.; Brooijmans, N. & Rizzo, R. C. (2006). Development and validation of a modular, extensible docking program: DOCK 5. *Journal of computer-aided molecular design, 20,* 601-619.

[33] Kontoyianni, M.; McClellan, L. M.; & Sokol, G. S. (2004). Evaluation of Docking Performance: Comparative Data on Docking Algorithms. *Journal of medicinal Chemistry, 47,* 558-565.

[34] Cross, S. S. J. (2005). Improved FlexX Docking Using FlexS-Determined Base Fragment Placement. *Journal of Chemical Information and Modeling, 45,* 993-1001.

[35] Verdonk, M. L.; Cole, J. C.; Hartshorn, M. J.; Murray, C. W. & Taylor, R. D. (2003). Improved Protein-Ligand Docking Using GOLD. *Proteins, 52,* 609-623.

[36] Ishii, N.; Robert, M. & Nakayama, Y. (2004). Toward large-scale modeling of the microbial cell for computer simulation. *Journal of Biotechnology, 113,* 281-294.

[37] Schmidt, H. & Jirstrand, M. (2006).Systems Biology Toolbox for MATLAB: a computational platform for research in systems biology. *Bioinformatics, 22,* 514 - 515.

[38] Weidemann, A.; Richter, S.; Stein, M.; Sahle, S.; Gauges, R.; Gabdoulline, R.; Surovtsova, I.; Semmelrock, N.; Besson, B.; Rojas, I.; Wade, R. & Kummer, U. (2008). SYCAMORE—a systems biology computational analysis and modeling research environment. *Bioinformatics, 24,* 1463 - 1464.

[39] Mirschel, S.; Steinmetz, K.; Rempel, M.; Ginkel, M. & Gilles, E. D. (2009). PROMOT: modular modeling for systems biology. *Bioinformatics, 25,* 687 - 689.

In: Recent Trends in Biotechnology, Volume 1
Editors: M.P. Singh, A. Agrawal and B. Sharma
ISBN 978-1-60876-148-7

Chapter 4

DNA BIOSENSOR : A QUICK MOLECULAR DIAGNOSTIC TECHNIQUE

***Ashok Kumar*[1] *and Suman*[2]**

[1] Institute of Genomics and Integrative Biology, Delhi University Campus
Delhi, India

[2] Institute of Advance Research and Studies (Material & Devices), Amity University
Noida, India

ABSTRACT

DNA based biosensors have recently gained much importance for detection of target genes responsible for diseases, in food industry, environment and in agriculture. Different types of DNA based biosensor, their advantages and basic principle of operating system has been described in this chapter. DNA biosensor provides fast, simple, economical, sensitive and selective detection of target genes by hybridization with specific probe. Various new strategies for DNA based biosensors have **been** described along with recent trends and challenges in development of technology. Electrochemical biosensor has more advantages due to the hybridization reaction on electrode surface or in solution in the presence of redox indicators. PCR free DNA biochip is emerging **as** new tools in the field of diagnosis.

INTRODUCTION

Biosensor is a quick analytical device and plays important roles in daily life. In last 20 years, biosensors have been increasingly used for continuous monitoring of biological and synthetic processes used in industrial and clinical chemistry. Biosensor is becoming popular in the field of food analysis [1], bioterrorism [2,3], environmental [4] and in the area of human health monitoring and diagnostics [5-7].There is vast exponential potential of biosensors. Presently, most fascinating and prospective sensors are immunosensors [8,9] based on affinity reactions between antibody and antigen and DNA sensors [10-16] based on hybridization of complementary ssDNA oligonucleotides.

In general, biosensor is small device employing biological recognition properties for a selective bioanalysis. Such devices rely on the intimate coupling of a biological recognition element with a physical transducer to convert the biological signals into an electrical signal or other signals, proportional to the concentration of analytes [17]. Biosensors eliminate the need of the sample preparation and hence offer great promise for several on-site analytical applications of rapid and cost effective measurements.

A basic biosensor assembly includes a receptor, transducer and processor. The sensing elements may be whole cells, antibodies, enzymes or nucleic acids forming a recognition layer that is integrated with transducer via immobilization by adsorption, cross-linking or covalent binding. The transducers are based upon the parameters of measurement. It may be amperometric (current measurement at constant potential) [18], potentiometric (potential measurement at constant current) [19], piezoelectric (measurement of changes in mass)[20], thermal (measurement of changes in temperature)[21] or optical (detect changes in transmission of light)[22].The usual analytical techniques require a number of steps, time and expensive instruments whereas biosensors are quick, simple, economical and may be used in small hospitals and laboratories of remote areas where sophisticated instrumental facilities are not available.

DNA Biosensor

The detection of specific DNA sequence is of significance in many areas including clinical, environmental and food analysis [4, 7, 23, 24]. The analysis of gene sequences and the study of gene polymorphisms play a fundamental role in rapid detection of genetic mutations, offering the possibility of performing reliable diagnosis even before any symptoms of a disease appear. In environmental and food areas the detection of specific DNA sequences can be used for the detection of pathogenic bacteria or genetically modified organism (GMO).

DNA biosensors and gene chips are of major interest due to their tremendous promise for obtaining sequence-specific information in a faster, simpler and cheaper manner compared to the traditional hybridization [25, 26].Recent advances in developing such devices opens a new opportunities for DNA diagnostics. DNA biosensors, based on nucleic acid recognition processes, are rapidly being developed towards the assay of rapid, simple and economical testing of genetic and infectious diseases. DNA sensors can be made by immobilizing single stranded (ss) DNA probes on different electrodes using electroactive indicators to measure the hybridization between DNA probes and their complementary DNA strands [27,28].

The current method for the identification of specific DNA sequence in biological samples are based on isolation of double stranded (ds) genomic DNA and further polymerase chain reaction (PCR) to amplify the target sequence of DNA. The PCR products can be subjected to electrophoresis or may be adsorbed onto a suitable membrane and exposed to a solution containing DNA probe (Southern Blot). The DNA probe is either chemically or enzymatically labeled with radioactive material, chemilumnophore or ligands such as biotin because nucleic acid itself is not able to provide any signal. Recent advances in the field of biomolecular techniques can be used to fabricate new generation miniaturized biosensor. The advantage and disadvantages of different types of DNA biosensors are summarized in Table 1.

1. Optical DNA Biosensors

Optical methods are the most frequently used in detection of analytes. The simplest detection based on spectrophotometer and fluorometers, which can be used for spectroscopic or fluorescence detection. Since nucleic acids do not have intrinsic properties that are functional in direct detection, many of the nucleic acid-based assays, especially optical setups, require a label for detection. The choice of label is based on stability, sensitivity and its availability.

Table 1. Advantage and disadvantage of different types of DNA biosensors

Type	Principle	Advantage	Disadvantages
1. Optical			
a Fiber optics	Evanescent wave based, allows measurement of binding at the fiber surface	Remote in-situ measurement, inherent sensitivity of optical approaches	Costly equipment and not portable
b. Laser interferometry	Planar waveguides have evanescent field causing change in refraction index	Highly sensitive, detect upto 1 cell	Susceptibility to turbidity interference
2. Electrochemical a. Conductometric b. Potentiometric c. Amperometric	Change in conductance Electric potential Oxidation/reduction	Fast, low cost	Highly buffered solution may interfere
3.Piezoelectric	Quartz crystals oscillation at fixed frequency, binding of an analyte causes change in oscillation frequency	Fast and highly sensitive	Sensitivity upto 1 cell is not demonstrated
4. Colorimetric/ Strip	Color development	Not required any instruments	Not quantitative
5. DNA biochip	Array based	Instrument required	Quantitative

DNA optical biosensors are based on a fiber optic to transduce the emission signal of a fluorescent label. Fiber optics are devices that carry light from one place to another by a series of internal reflections [29]. The operation of fiber-optic DNA biosensors involves placement of a ssDNA probe at the end of the fiber and monitoring the fluorescent changes resulting from the association of a fluorescent indicator with the double-stranded (ds) DNA hybrid [30-31]. The first DNA optical biosensor, developed by Krull and coworkers using fluorescent indicator ethidium bromide [30,32].Watts group developed a fiber-optic DNA sensor array for the simultaneous detection of multiple DNA sequences [33].The hybridization of fluorescent labeled complementary oligonucleotides was monitored by observing the increase in fluorescence.

A different optical transduction, based on evanescent wave devices, can offer real-time label-free optical detection of DNA hybridization. The different types of optical biosensors are as follows:

1.1. Molecular Beacons (MBs)

MBs are oligonucleotides with a stem-and-loop structure, labeled with a fluorophore at one end and a quencher on the other end of the stem that become fluorescent upon hybridization (Figure 1).In addition to their direct monitoring capability, MB probes offer high sensitivity and specificity. A biotinylated molecular beacon probe was developed to prepare a DNA biosensor using a bridge structure. MB was biotinylated at quencher site of the stem and linked on a biotinylated glass through avidin or streptavidin, which acted as bridge between MB and glass matrix. The fluorescence change was measured by confirmation change of MB in the presence of complementary target DNA [34, 36].

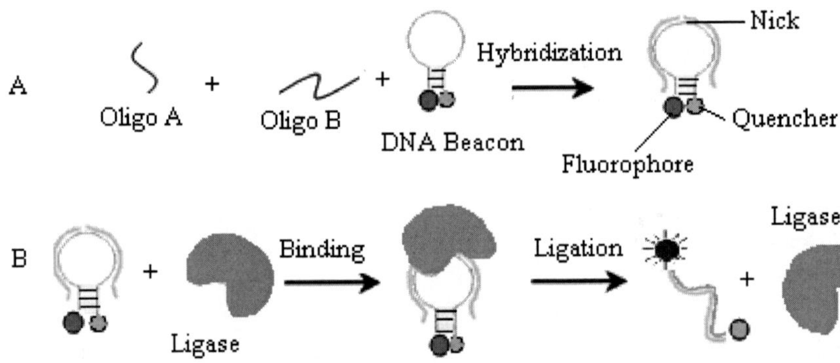

Figure 1. DNA biosensor based on Molecular Beacons.

1.2. Surface Plasmon Resonance (SPR)

Surface plasmon resonance (SPR) is a quantum optical-electrical phenomenon arising from the interaction of light with metal surface (Figure 2). Under specific conditions the energy carried by photons of light is transferred to packets of electrons (photons) on a metal surface. Energy transfer occurs only at specific resonance wavelength of light. These biosensors are based on monitoring the changes in the surface optical properties (change in resonance angle due to change in the interfacial refractive index) resulting from the surface binding reaction. Such devices thus combine the simplicity of SPR with the sensitivity of wave guiding devices. The resonance conditions are influenced by the material adsorbed onto the thin metal film. A linear relationship is found between resonance energy and mass concentration of molecules such as proteins, sugars and DNA. The SPR signal which is expressed in resonance units is therefore a measure of mass concentration at the sensor chip surface [37,38].It has been reported that alkane thiol modified oligonucleotide can be immobilized onto gold surface to detect DNA hybridization using SPR based detection in agro food industry [39].

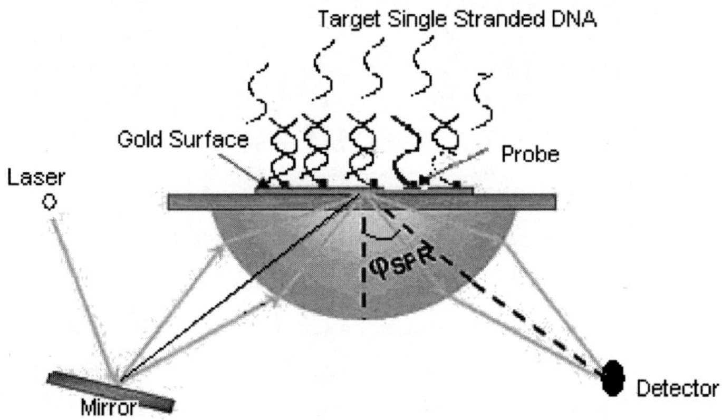

Figure 2. Surface Plasmon Resonance (SPR) based DNA biosensor.

1.3. Quantum–Dot

An ultrasensitive nanosensor based on fluorescence resonance energy transfer (FRET) can detect very low concentration of DNA and do not require separation of unhybridized DNA. Such type of technique is based on quantum-dots (QDs) which are linked to specific DNA probes to capture target DNA (Figure 3). The target DNA strand binds to a fluorescent-dye (fluorophore) labeled reporter strand and thus forming FRET donor-acceptor assembly. Quantum dot also functions as to concentrate the signal by confining several targets at nanoscale domain.

Unbound DNA strand produce no fluorescence but on binding of even small amount of target DNA ($\leq$ 50 copies) may produce very strong FRET signal. Several FRET based DNA probes (molecular beacons and TaqMan probes) whose fluorescence signals change as a result of hybridization or enzymatic reactions have been developed for separation free (unhybridized DNA strand) detection of target DNA [40-42].

DNA nanosensor consists of two target specific DNA probes i.e. reporter and capture probe. The reporter probe is labeled with fluorophore whereas capture probe is labeled with biotin which binds with avidin conjugated with QD. The QD functions as target concentrator as well as FRET energy donor. When target DNA is present in solution, it becomes sandwiched by reporter and capture probes. Several sandwiched hybrids are then captured by a single QD through biotin-avidin binding and concentrate at nanoscale domain [42].The fluorophore acceptor and QD donor close proximity causes fluorescence from acceptor by means of FRET on illumination of the donor. The detection of acceptor emission indicates the presence of target DNA. The unhybridized probe do not participate in FRET and do not give fluorescence, therefore, it is not necessary to remove. [42].

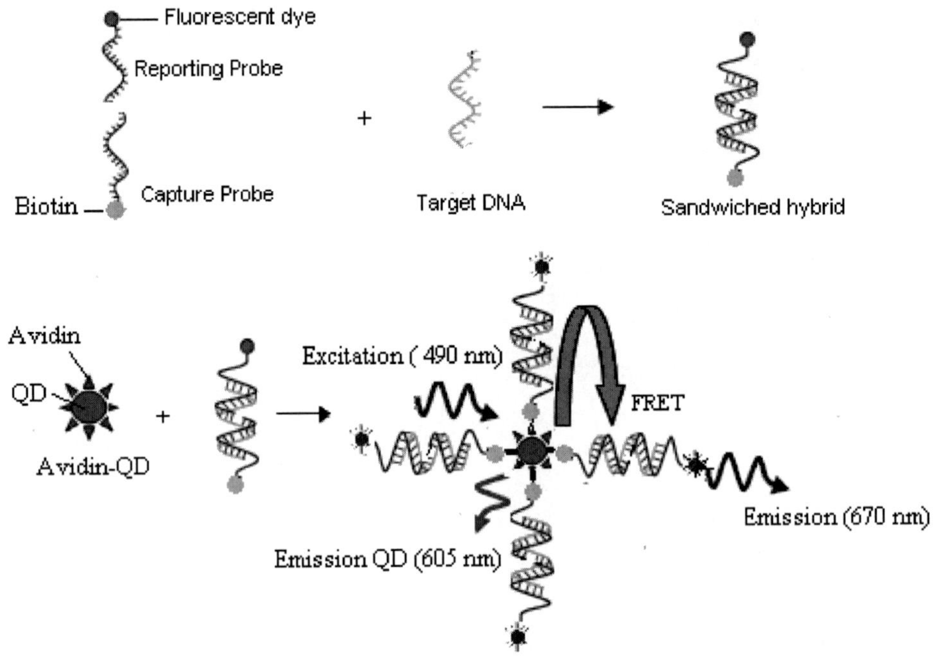

Figure 3. Quantum-Dot based DNA biosensor.Fluorescence emission from fluorescent dye illumination at QD caused by FRET between fluorescent dye acceptors and QD donor.

The fluorescent dyes used as the standard labels for this type of optical biosensors are very expensive and it can rapidly photo-bleach (the dye is photochemically converted to a non-fluorescent compound). An alternative to the fluorescence detection used in many systems is chemiluminescence based, which overcomes the use of fluorescent dyes.

2. Electrochemical DNA Biosensors

Electrochemical devices are very useful for sequence-specific biosensing of DNA. The miniaturization of devices and advanced technology make them excellent tool for DNA diagnostics. Electrochemical detection of DNA hybridization usually involves monitoring a current at fixed potential. Electrical modes were developed for detection of both label-free and labeled objects [43-50].The immobilization of the nucleic acid probe onto the transducer surface plays an important role in the overall performance of DNA biosensors and gene chips [51-53].

The immobilization step requires a well-defined probe orientation and accessible to the target for hybridization. Depending upon the nature of the physical transducer, various methods can be used for attaching the DNA probe to the solid surface such as the use of thiolated DNA probe for self assembled monolayers (SEM) onto gold transducers by covalent linkage to the gold surface via functional alkanethiol-based monolayers (Figure 4).

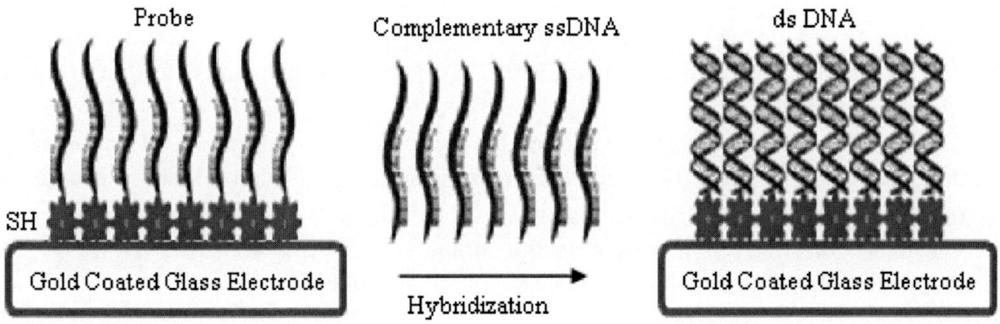

Figure 4. Schematic representation of immobilized thiolated (-SH) DNA probe onto gold surface and hybridization of complementary DNA (ssDNA) with probe.

The other method of attachment of DNA probe is to biotinylate DNA probe and attachment through biotin-streptavidin interaction on electrode surface (Figure 5) [12, 45, 54].The avidin modified polyaniline electrochemically deposited onto a Pt disc electrode for direct detection of *E.coli* **by immobilizing a 5' biotin labeled probe using a differential pulse** voltametric technique in the presence of methylene blue as a DNA hybridization indicator [14, 44]. Similarly, electrochemical DNA biosensor based on polypyrrole-polyvinyl sulfonate coated onto Pt disc electrode was also fabricated using biotin-avidin binding [45].

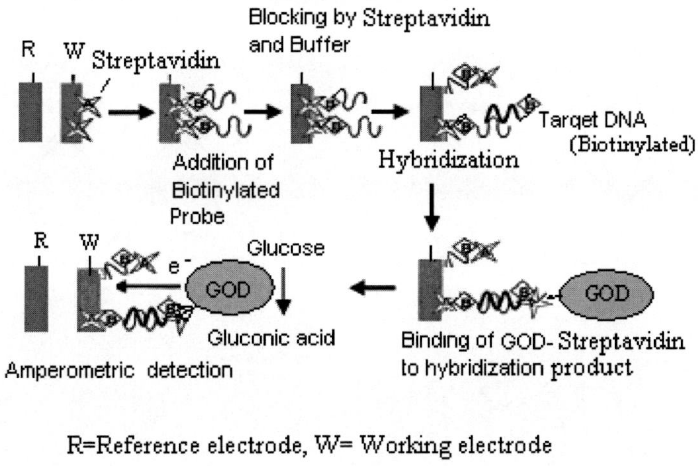

Figure 5. Amperometric DNA biosensor using biotin-streptavidin interaction on electrode surface.Oxidation of glucose to gluconic acid by glucose oxidase causing change in the current.

The discovery of carbon nanotubes (CNTs) in DNA analysis plays an important role for development of electrochemical DNA biosensor. CNT not only enables immobilization of DNA molecules but is also used as powerful amplifier to amplify signal transduction of hybridization. CNT also works as novel indicator of hybridization. The application of arrayed CNT into DNA chip require small amount of sample. The CNT based biosensor play major role on DNA based diagnostics in hospitals or at home [55].The knowledge of peptide nucleic acid (PNA) has opened a new research area of DNA biosensors. PNA is a DNA mimic in which the sugar phosphate backbone is replaced with a pseudopeptide. The unique structural, hybridization and recognition features of solution-phase PNA can be readily extrapolated onto transducer surfaces in connection with the design of highly-selective DNA biosensors. PNA has remarkable sequence specificity onto DNA biosensors including detection of single-base mismatches [53].

The hybridization is commonly detected by the increase in current signal due to redox indicator (that recognizes the DNA duplex) or from other hybridization-induced changes in electrochemical parameters (e.g. conductivity or capacitance). New redox indicators are offering greater discrimination between single strand (ss) and dsDNA [27, 28, 47, 56]. The electrochemical DNA biosensor may be labeled based and labeled free.

2.1. Label Based or Indirect Detection

In label based electrochemical biosensor specific organic dyes, metal complexes or enzymes are used for hybridization detection. The use of enzyme labeled probe offers great promise for electrochemical detection of DNA hybridization. On addition of substrate to the enzyme modified electrode surface, the electrochemical activity of the product simplifies the detection of hybridization [48]. Redox- active molecules such as daunomycin, meloda blue and methylene blue which is inserted between the dsDNA and gives signal which can be used for hybridization detection (Figure 6) [48]. Redox-active molecules based two commercialized DNA chips have been introduced in molecular diagnosis market in the trade name of eSensor™ produced by Motorola Life Sciences [25], Inc.and Genlyser ™ by Toshiba [26].

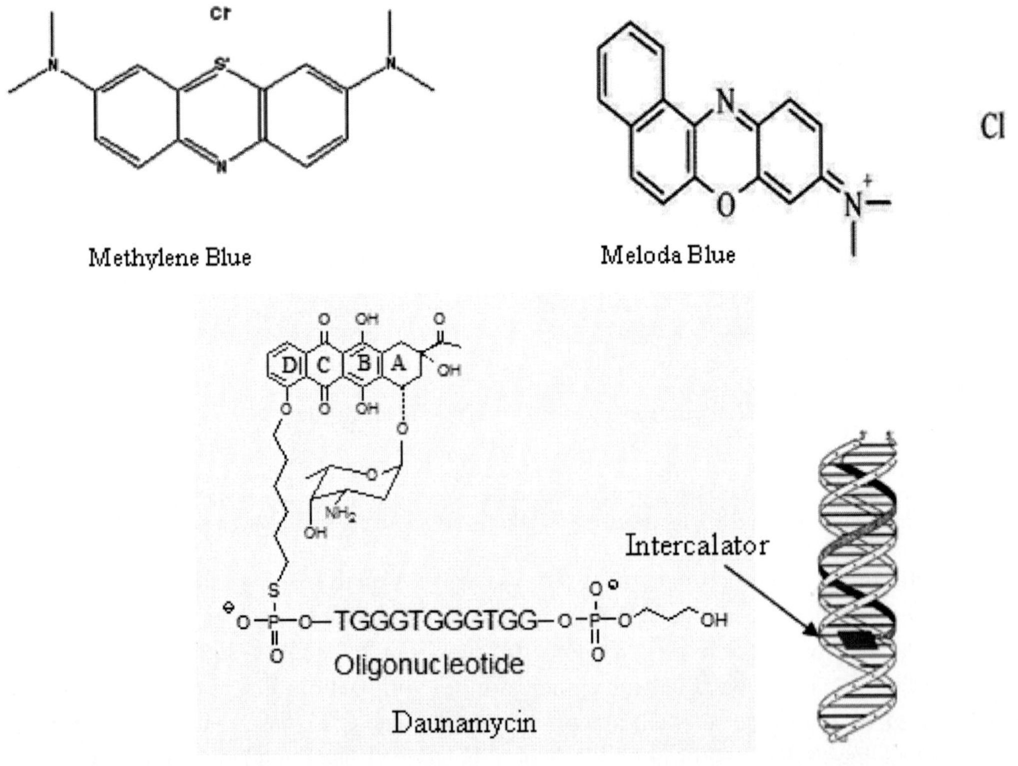

Figure 6. Structure of redox intercalating indicators used for hybridization detection.

2.2. Label Free or Direct Detection

In direct detection techniques, a target molecule or any other object from the system does not need to be labeled [48].Although label-dependent methods achieve the highest sensitivities, eliminating the labeling steps simplifies the readout, the speed and ease of nucleic acid assays.

In a label-free method the immobilized probe recognizes a complementary sequence if the target is present in the sample. Next, the transducer converts the biological interaction into a measurable signal, proportional to the degree of hybridization that is to the amount of target molecule in the sample. Label-free strategies reduce analysis times and cost. They are also free from unfavorable effects from the labels, such as instability and steric hindrances.

Recently, a new label-free electrochemical detection technique has been developed which is faster and simpler [14, 15, 43, 45]. It is possible to exploit changes in the intrinsic electroactivity of DNA (guanine oxidation peak of hybridization). To overcome the limitations of the probe sequences (absence of G), guanines in the probe sequence were substituted by inosine residues (pairing with C) and the hybridization was detected through the target DNA guanine signal [15].Direct, label-free, electrical detection of DNA hybridization has also been accomplished by monitoring changes in the conductivity of conducting polymer molecular interfaces (DNA-modified polypyrrole films).

3. Piezoelectric DNA Biosensor

Piezoelectric DNA biosensor is based on quartz crystal that oscillate at a defined frequency at applied oscillating voltage. Piezoelectric method have recently emerged as most attractive due to their simplicity, cost, sensitivity and real time label-free detection [20, 57].

The quartz crystal microbalance (QCM) is an extremely sensitive mass-measuring device that allows dynamic monitoring of hybridization events. QCM hybridization biosensors consist of an oscillating crystal with the DNA probe immobilized on its surface. The increased mass, associated with the hybridization reaction, results in a decrease of the oscillating frequency (Figure 7). QCM transducers also used for investigating DNA interactions, including real-time detection of protein-DNA binding or monitoring of enzymatic cleavage reactions.

Oligonucleotide QCM sensor for the microalgae *Alexandrium minutum* was developed by immobilizing probe complementary strand of a partial sequence of the gene encoding microalgae that produces neurotoxins responsible for paralytic shellfish poisoning on European and Asian coasts [58]. After hybridization *in situ* by using a 27 MHz quartz crystal microbalance the frequency changes under controlled hydrodynamic conditions. Piezoelectric biosensor was also developed for *Salmonella typhimurium* [59].

Genomic DNA of *E.coli* hybridizes with the same rate constant on the QCM biosensor as in homogeneous solution. A high hybridization rate was obtained when nucleic acids are hybridized in a thin film, micro volume reaction on a non porous surface [60]. A DNA piezoelectric sensor was also developed for detection of genetically modified organisms (GMO) by immobilizing. DNA probe on the sensor surface of a QCM device and hybridization of probe with target DNA was monitored in solution. The above technique is sensitive and specific for detection of GMO and provides a useful tool for screening and analysis of food [64].

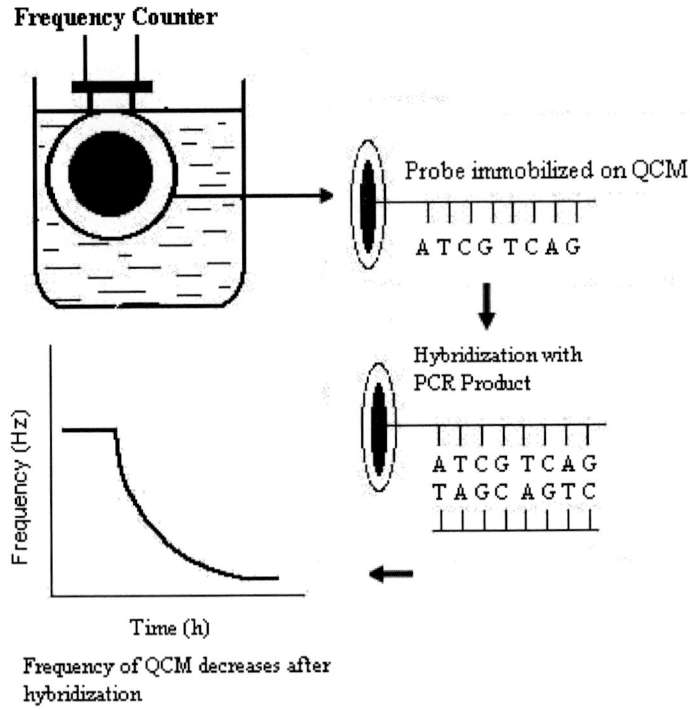

Figure 7. Rapid detection of specific PCR product by hybridization on immobilized DNA probe at quartz crystal microbalance.

The piezoelectric sensor can be used for genetically modified soybean without PCR amplification [62].The hybridization of probes with complementary, non complementary and mismatched DNA was able to distinguish polymorphism.

4. Colorimetric or Strip type DNA Sensor

A novel nanoparticle based colorimetric detection offers great promise for direct detection of DNA hybridization [63-65]. In this case, a distance change, occurred from the hybridization event, results in changes of the optical properties of the aggregated functional gold nanoparticles. The dry-reagent strip type biosensor has been developed for visual detection of DNA within a short time [66]. Oligonucleotide conjugated gold nanoparticle is used as probe for detection of target DNA through hybridization (Figure 8). The advantage of this type of biosensor is that it does not require any instruments, multiple incubation and washing steps as performed in most assays. Gold nanoparticle with oligo (dT) attached to their surface form integral part of the strip. Biotinylated PCR products are hybridized with poly (dA) tailed oligo and applied on the strip and immersed in the appropriate buffer. As the buffer migrates upward, it rehydrates the nanoparticle that are linked through target DNA through poly (dA/dT) hybridization. The hybrid is then captured by immobilized avidin or strepatavidin in the test zone of the strip and generate red band. Another, red band is formed by hybridization in the control zone of the strip to indicate proper test performance [66]. The test is 8-10 times more sensitive than ethidium bromide in agarose gel electrophoresis. The detection limit is as low as 2 f mol of amplified DNA products.

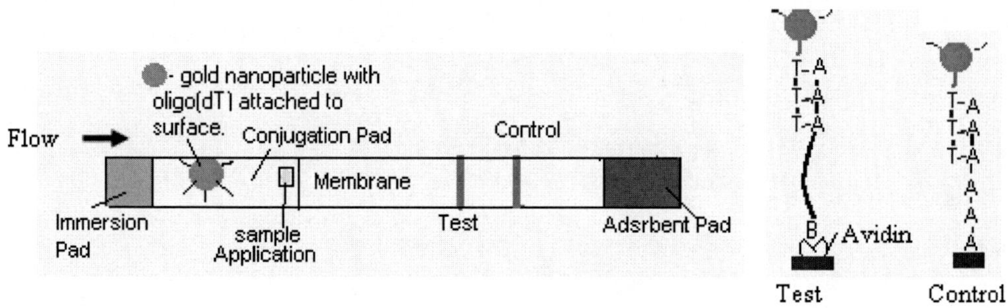

Figure 8. DNA biostrip and formation of colour on test and control zone on hybridization of target DNA.

4. DNA Biochip

Microarrays, DNA arrays, gene chips or biochips are same terminology often being intermixed to describe analysis of complex DNA samples and expression of genes [52, 53, 67-70].The most attractive features of the biochip is the miniaturization, fastness and accuracy. The DNA microchip technology offers an enormous potential for fast multiplex analysis of nucleic acid samples, diagnosis of genetic diseases, detection of infectious agents, measurements of gene expression, drug screening and forensic analysis.

The biochips are fabricated from glass, silicon or plastic supports and comprise thousands of 10-100 μm reaction zones onto which individual oligonucleotides have been deposited. Biochip has high densities (upto 10^6 sites/cm^2) in connection with typical 1-2 cm^2-size chips. The exact number of probes varies in with the application. The construction of gene chips involves the immobilization or synthesis of an array of DNA probes on a solid support. High-density DNA arrays often require the use of microjet deposition technique (picoliter volumes) onto discrete locations on the chip. It is essential to activate the surface for a covalent attachment of the oligonucleotide probes.

DNA chip technology requires development of methods for fabrication of the probe arrays, target hybridization, algorithms for analyzing the data and reconstructing the target sequence. The automation of gene chip greatly facilitates production and accelerates their operation, while eliminating human errors.The detection of the DNA hybridization (at individual spots) relies on the signal generated by the binding event [71-73].

The most common application of DNA/oligonucleotide microarray is gene expression analysis (Figure 9). In this technique, RNA isolated from two samples are labeled with two different fluorochromes (generally the green cyanine 3 and the red cyanine 5 (Cy3, Cy5)) before being hybridised to a microarray consisting of large numbers of cDNAs / oligonucleotides orderly arranged onto a glass microscope slide. After hybridization under stringent conditions, a scanner records, after excitation of the two fluorochromes at given wavelengths, the intensity of the fluorescence emission signals that is proportional to transcript levels in the biological samples. The microarray data are analysed by specific softwares.

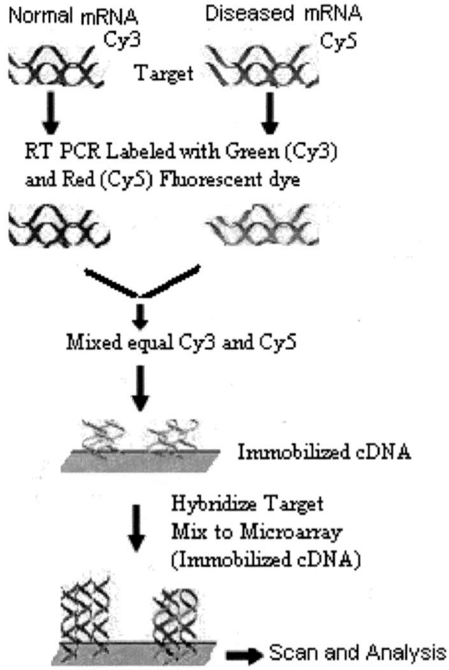

Figure 9. Schematic representation of DNA biochip (Microarray).

5. Future Prospects

The use of DNA biostrip and biochip technologies eliminates the role of PCR. Future biosensors will require the development of new reliable devices and the improvement of the existing sensors for better transduction, amplification and conversion of the biological signals. Efficient biosensors may not function as a stand-alone detector, but will form an integral part of an analytical system. Compact and portable devices will constitute bright future area of multidisciplinary sensor research. Further, increase of interest on DNA based sensors can be expected in near future together with a commercial production of these devices and their wide use. However, basic research is still necessary to improve the sensor technologies, sensing strategies as well as analytical instrumentations and procedures [74-75].

Conclusion

Different types of electrodes immobilized with specific probes can be used to detect the presence of complementary target sequence by hybridization technique. Besides the different immobilization methods, electroactive hybridization indicators and different conducting polymer based on nano-composites are also used for development of electrochemical biosensors. SPR, Quantum-Dot and piezoelectric biosensors are the emerging area of molecular diagnosis. The Opto sensors interfacing based on universal frequency-to-digital converter and biochip have opened new opportunities for development of DNA biosensors.

REFERENCES

[1] Eden-Firstenberg, R., Schaertel, B. J.(1988) Biosensor in the food industry: Present and Future, *J. of Food Protection*, 51, pp.811-820.

[2] Lindner, D. (2001) R&D at Scandia National Laboratories, *Lab as a chip*, 1, pp.15-19.

[3] Burkle, F. M.(2003) Measures of effectiveness in large scale bioterrorism events, *Prehosp Disast. Med.,*18, pp.258-262.

[4] Mascini, M. (2001) Affinity electrochemical biosensors for pollution control, *Pure Appl. Chem.,* 73, pp.23-30.

[5] Malhotra, B. D., Chaube, A.(2003) Biosensors for clinical diagnostics industry, *Sensor Actuators B,* 91, pp.117-126.

[6] Annelies, B., Dirk, B., Michel, L.(2005) Clinical and analytical performance of the Accu-chek inform point-of-care glucose water, *Point of Care*, 4, pp.36-40.

[7] Anjum, V. & Pundir, C. S.(2007) Biosensors: Future analytical tools, *Sensors and Transducers*, 76, pp.937-944.

[8] Blonder, R., Levi, S., Tao, G., Ben-Dov, I., Willner, J.(1997) Development of amperometric and microgravimetric immunosensors and reversible immunosensors using antigen and photoisomerizable antigen monolayer electrodes, *J. Am. Chem. Soc.,*119, pp.467- 478.

[9] Blonder, R., BenDov, I., Dagana, A., Willner, I., Zisman, E (1997) Photochemically activated electrodes: application in design of reversible immunosensors and antibody patterned interfaces, *Biosens. Bioelectron.,* 12, pp.627-644.

[10] Prummond, T. G.(2003) Electrochemical DNA Sensors, *Nature Biotechnol,.* 21, pp.1192-1199.

[11] Umek, R. M.(2001) Electronic detection of nucleic acids, *J. Mol. Diagnostics*,3, pp.74-84.

[12] Junhui, Z., Hong, C., Ruifu, Y. (1997) DNA based biosensors, *Biotech. Advanc.,*15, pp.43-58.

[13] Arora, K., Chand, S., Malhotra, B. D.(2006) Recent developments in biomolecular electronics techniques for food pathogens, *Anal. Chim Acta,* 568, pp.259-274.

[14] Arora, K.,Prabhakar, N.,Chand, S., Malhotra, B.D.(2007) Ultrasensitive DNA hybridization biosensor based on polyaniline, *Biosens. Bioelectron.,23,* pp.613-620.

[15] Prabhakar, N., Arora, K., Singh, S. P., Pandey, M. K., Singh, H., Malhotra, B. D.(2007) Polypyrrole-polyvinyl sulphonate film based disposable nucleic acid biosensor, *Anal. Chim Acta,* 589, pp.6-13.

[16] Berdat, D., Marin, A., Herrera, F., Gijs, M. A.M.(2006) DNA biosensors using fluorescence microscopy and impedance spectroscopy, *Sensors Actuators*, 118, pp.53-59.

[17] Sharma, S. K., Sehgal, N., Kumar, A.(2003) Biomolecules for development of biosensors and their application, *Curr. Appl. Phys.,* 3, pp.307-316.

[18] Ho, K.C., Cheu, C. Y., Hsu, H. C., Cheu, L. C., Shiesh, S. C., Z.lin, X. (2004) Amperometric detection of morphine at a prussian blue modified indium tin oxide electrode, *Biosens. Bioelectron.,* 20, pp.3-8.
[19] Wang, J., Xu, D.K., Kawde, A. N., Polsky, R.(2001) Metal nanoparticle-based electrochemical stripping potentiometric detection of DNA hybridization, *Anal. Chem.,*73, pp.5576–5581.
[20] Bunde, R. L., Jarvi, E. J., Rosentreter, J. J.(1998) Piezoelectric quartz crystal biosensors, *Talanta,* 46(6), pp. 1223-1236.
[21] Xie, B. (2000)Mini/micro thermal biosensors and other related devices for biochemical / Clinical monitoring, *TRAC,* 19, pp.340-347.
[22] Mebravar, M.(2000)Fiber optic biosensors:Trends and advances, *Anal Sci.,*16, pp.677-692.
[23] Ivniski, D., Hamid, I. A., Atanasov, P., Wilkins, E.(1999) Biosensors for detection of pathogenic bacteria, *Biosens. Bioelectron.*, 14, pp. 559-624.
[24] Berney, H., West, J., Haefele, E., Alderman, J., Lane, W., Collins, J. K.(2000) DNA diagnostic biosensor development, characterization and performance, *Sens. Actuators B*, 68, pp.100-108.
[25] Motorola Life Sciences Inc. (www.motorola.com/Lifesciences).
[26] Toshiba Corporation(www.dna-chip.toshiba.co.jp/eng/)
[27] Erdem, A., Kesman, K., Mesie, B., Akarea, U. S., Ozsoz, M.(2000) Novel hybridization indicator methylene blue for the electrochemical detection of short DNA sequence related to Hepatitis B Virus, *Anal. Chim. Acta*, 423, pp.139-149.
[28] Millan, K. M. and Mikkelsen, S. R.(1993) Sequence selection biosensor for DNA based on electroactive hybridization indicators, *Anal. Chem.*, 65, pp. 2317-2324.
[29] Byoungho, L.(2003) Review of the present status of optical fiber sensors, *Optical Fiber Technology*, 9(2), pp.57-79.
[30] Piunno, P.A.E., Krull, U. J., Hudson, R. H. E., Damha, M. J. and Cohen, H.(1994) Fiber optic biosensor for fluorometric detection of DNA hybridization, *Anal. Chim. Acta,* 288, pp. 205-214.
[31] Piunno, P.A.E., Krull, U. J., Hudson, R.H.E., Damha, M. J., Cohen, H.(1995) Fiber optic DNA sensor for fluorometric nucleic acid determination, *Anal. Chem.,* 67, pp. 2635-2643.
[32] Haughland, R. P.(1992) Molecular Probes: Handbook of fluorescent probes and research chemicals, 5th ed Molecular Probes.
[33] Watts, H. J., Yeung, D., Parkes, H.(1995) Real time detection and quantification of DNA hybridization by an optical biosensor, *Anal. Chem.,* 67, pp. 4283-4289.
[34] Weihong, J. L., Wang, K., Xiao, D., Yang, X., He, X., Tang, Z.(2001)Ultrasensitive optical DNA biosensor based on surface immobilization of molecular beacon by a bridge structure, *Anal. Sci.*, 17, pp. 1149-1153.
[35] Cady, N.C., Strickland, A. D., Batt, C. A.(2007)Optimized linkage and quenching strategies for quantum dot molecular beacons, *Mol. Cell. Probes,* 21, pp. 116-124.
[36] Liu, X., Farmerie, W., Schuster, S., Tan, W.(2000) Molecular beacons for DNA biosensors with micrometer to submicrometer dimensions, *Anal. Biochem.,* 283, pp.56-63.

[37] Wang, R., Tombelli, S., Minunni, M., Michelia, M. Spiriti, M.(2004) Immobilization of DNA probes for the development of SPR- based sensing, *Biosens. Bioelectron.,* 20, pp. 967-974.

[38] 38]. Yu, F.,Yao, D.,Knoll,W.(2004) Oligonucleotide hybridization studied by a surface plasmon diffraction sensor (SPDS), *Nucl.Acid .Res.,* 32(9), pp 1-7.

[39] Rella, R., Spadavecchia, J., Manera, M. G., Sciliano, P., Santino, A., Mita,G, (2004) Surface plasmon resonance imaging technique for nucleic acid detection, *Biosens. Bioelectron.,* 20, pp.1140-1148.

[40] Han, X., Gao, J., Su, Z., Nie, S.(2001) Quantum-dot tagged microbeads for multiplexed optical coding of biomolecules, *Nature Publishing Group*, 19, pp 631-635.

[41] Wang, H. J., Kawde, A. N., Xiang, Y., Gothelf, V., Collins, G.(2006) Quantum dot aptamer based ultrasensitive multi analyte electrochemical biosensor, *J. Am. Chem. Soc.,*128(7), pp.2228-2229.

[42] Chun-Yang Zhang, Hsin-Chih Yeh, Kuroki, M.,Tza-Huei Wang (2005) Single quantum-dot-based DNA nanosensor, *Nature Materials*, 4, pp. 826-830.

[43] Prabhakar, N., Arora, K., Singh, S. P., Singh, H., Malhotra, B. D.(2007) DNA entrapped polypyrrole-polyvinyl sulfonate film for application to electrochemical biosensor, *Anal. Biochem.*,366, pp.71-79.

[44] Arora, K., Prabhakar, N., Chand, S., Malhotra, B. D.(2007) *Escherichia coli* Genosensor Based on Polyaniline, *Anal. Chem.*, 79, pp. 6152-6158.

[45] Arora, K., Prabhakar, N., Chand, S., Malhotra, B. D.(2007) Immobilization of single stranded DNA probe onto polypyrrole-polyvinyl sulfonate for application to DNA hybridization biosensor, *Sensors Actuators B*,126, pp.655-663.

[46] Kara, P., Kerman, K., Ozkan, D., Meric, B., Erdem, A., Ozkan, Z., Ozsoz, M.(2002) Electrochemical genosensor for the detection of interaction between methylene blue and DNA, *Electrochem. Commun.,* 4, pp.705-709.

[47] Erdem, A., Kerman, K., Meric, B., Ozsoz, M.(2001) Methylene blue a novel electrochemical hybridization indicator, *Electroanalysis*, 13(3), pp.219-223.

[48] Kobayashi, K. M., Tamiya, E.(2004) Recent trends in electrochemical DNA biosensor technology, *Meas. Sci. Technol.*, 15, pp.1-11.

[49] Cheng, G., Zhao, J., Tu, Y., He, P., Fang, Y.(2005) A sensitive DNA electrochemical biosensor based on magnetite with a glassy carbon electrode modified by multi-walled carbon nanotubes in polypyrrole, *Anal Chim Acta,* 533, pp.11-16.

[50] Mascini, M., Palchetti, I., Murazza, G. (2001) DNA electrochemical biosensor, *Fresenius J. Anal Chem.,* 369, pp.15-22.

[51] Chrisey, L. A., Lee, G. U., Ferrall, C. E. O.(1996) Covalent attachment of synthetic DNA to self-assembled monolayer films, *Nucl. Acids Res.*, 24(15), pp.3031-3039.

[52] Marks, R. S., Lowe, C. R.,Cullen, D. C.,Weetal, H. H., Karube, I.(2007) *Biosens. Biochips*, Willey Publishers.

[53] Wang, J. (2008)From DNA biosensors to gene chips, *Nucl .Acid.Res.,* 28(16), pp 3011-3016.

[54] Marrazza, G., Chinella, I., Mascini, M.(1999) Disposable DNA electrochemical sensor for hybridization detection, *Biosens. Bioelectron.*, 14, pp. 43-51.

[55] Pingang, H., Ying, X., Yuzhi, F.(2006) Application of carbon nanotubes in electrochemical DNA biosensor (Review), *Microchim Acta*, 152, pp. 175-186.

[56] Yang,W., Ozsoz, M.,Hibbert, D.B.,Gooding, J.J.(2002) Evidence for the direct interaction between methylene blue and guanine bases using DNA-modified carbon paste electrodes, *Electroanalysis,*14 (18), pp.1299-1302.

[57] Kumar, A.(2000)Biosensors based on piezoelectric crystal detectors: Theory and applications, *JOM*, 52 (10), pp. 1-10.

[58] Lazerges, M., Perrot, H., Antoine, E., Defontaine, A., Compere, C.(2006) Oligonucleotide quartz crystal microbalance sensor for the microalgae *Alexandrium minutum* (Dinophyceae), *Biosens. Bioelectron.,* 21, pp. 1355-1358.

[59] Ye,M.,Letcher, S.V.,Rand, A.G.(1997) Piezoelectric biosensor for detection of *Salmonella typhimurium*, J. Food Sci., 62(5), pp.1067-1076.

[60] Towery, R.B., Fawcett, N.C., Zhang, P., Evans, J. A.(2001) Genomic DNA hybridizes with the same rate constant on the QCM biosensor as in homogeneous solution, *Biosens. Bioelectron.*, 16, pp.1-8.

[61] Mannelli, I., Minunni, M., Tombelli, S., Mascini, M.(2003) Quartz crystal microbalance (QCM) affinity biosensor for genetically modified organisms (GMOs) detection, *Biosens. Bioelectron.,* 18, pp.129-140.

[62] Stobiecka, M., Ciesla, J. M., Janowska, B., Tudek, B., Radecka, H.(2007) Piezoelectric sensors for determination of genetically modified soybean roundup ready in samples not amplified by PCR, *Sensors,* 7, pp.1462-1459.

[63] Storhoff, J.J., Elghanian, R., Mucic, R. C., Mirkin, C.A., Letsinger, R.L.(1998) One-pot colorimetric differentiation of polynucleotides with single base imperfections using gold nanoparticle probes, *J. Am. Chem. Soc.,* 120, pp.1959-1964.

[64] Juewen Liu, Yi Lu (2004) Calorimetric biosensors based on DNAzyme-assembled gold nanoparticles, *Fluores.,*14(4), pp.343-354.

[65] Sato,K.,Hosokawa, K.,Maeda, M.(2007)Colorimetric biosensor based on DNA nanoparticle conjugates, *Anal.Sci.,* 23, pp.17-20.

[66] Glynou, K., Loannou, P.C., Christopoulos, T.K., Syriopoulou, V.(2003) Oligonucleotide-functionalized gold nanoparticles as probes in a dry-reagent strip biosensor for DNA analysis by hybridization, *Anal Chem.*, 75, pp. 4155-4160.

[67] Taton, T. A., Mirkin, C. A., Letsinger, R. L. (2000) Scanometric DNA array detection with nanoparticle probes, *Science*, 289, pp.1757-1760.

[68] Ramsay, M. (1998) DNA chip: state-of-the art, *Nat. Biotechnol.,* 16, pp. 40-44.

[69] Peter, C., Meusel, M., Grawe, F., Katerkamp, A., Cammann, K.(2001) Optical DNA-sensor chip for real time detection of hybridization events, *Fresenius J. Anal. Chem.*, 371, pp.120-127.

[70] Ferguson, J.A., Boles, T. C., Adams, C. P., Walt, D. R.(1996) A fiber-optic DNA biosensor microarray for the analysis of gene expression, *Nat.Biotech.,*14, pp.1681-1684.

[71] Sethi, D., Kumar, A., Kumar, P., Gupta, K.C. (2008) A facile method for construction of oligonucleotide microarrays, *Bioconj. Chem.,*19, pp.2136-2143.

[72] Sethi, D., Patnaik, S, Kumar, A., Gandhi, R.P., Gupta, K.C., Kumar, P. (2009) Polymer supported synthesis of aminooxyalkylated oligonucleotides and some applications in the fabrication of microarrays, *Bioorg. Medicin. Chem.,*17, pp.5442-5450.

[73] Mahajan, S., Sethi, D., Seth, S., Kumar, A., Kumar, P., Gupta K.C.(2009) Construction of oligonucleotide microarray (Biochip) via thiolether linkage for the detection of bacterial meningitis, *Bioconj. Chem.(accepted available online).*

[74] Patel, M. K., Solanki, P., Seth, S. Khare, S., Gupta, S., Kumar, A., Malhotra, B. D. (2009) Ctr A gene based electrochemical DNA sensor for detection of meningitis, *Electrochem. Commun.* 11, pp.969-973.

[75] Suman and Kumar, A. (2008) Recent advances in DNA biosensors, *Sensor. Transdu. J.* 92(5), pp.122-133.

In: Recent Trends in Biotechnology, Volume 1
Editors: M.P. Singh, A. Agrawal and B. Sharma

ISBN 978-1-60876-148-7

Chapter 5

DNA FINGERPRINTING AND ITS APPLICATION

Varsha Srivastava*

Centre for DNA Fingerprinting and Diagnostics,
Tuljaguda, Nampally, Hyderabad-500 001, AP, India

ABSTRACT

DNA fingerprinting is a very powerful technology which is used by forensic scientists to establish identity. Such DNA-based technologies can help in the fight against crime. This technology has experienced many advancements occurring with the passage of time. Variable number tandem repeats (VNTR) are abundant in our genome and these regions are analyzed in DNA profiling.DNA testing is a matter of serious concern as it involves many ethical issues. DNA fingerprinting can be a boon to our society in different walks of our life if used efficiently.

INTRODUCTION

DNA fingerprinting is a powerful technology that has revolutionized forensic science. No two individuals can have an identical DNA pattern. Only exception is identical twins. Such DNA-based technologies have enormous social implications and especially in fight against the crime.DNA technology was first developed in England in 1985 by Sir Alec Jeffreys using restriction fragment length polymorphism (RFLP). DNA fingerprinting technology, named so because DNA is used for identification rather than fingerprints, gives a unique and specific profile of an individual just like thumb impression [13]. About 95% of the human genome is non-coding and the rest less than 5% of the genome is transcribed to protein. These non-coding regions, called Junk DNA, are abundant in repetitive sequences. The number of repeats shows individual variation. These DNA polymorphisms change the length of the DNA fragments produced by the digestion of restriction enzymes.

* Tel: 91-40-24749321/22/23; Fax:91-40-24749448; Email: varsha@cdfd.org.in

A DNA fingerprint is constructed as follows: The first step is extracting DNA from the sample. The DNA is then cut at specific sites by restriction enzymes, and the segments are arranged as per their molecular size using electrophoresis. The segments are marked with probes and exposed on X-ray film where they form a characteristic pattern of black bars—the DNA fingerprint. The exact number and size of fragments produced by a specific restriction enzyme digestion varies and is a characteristic of an individual. When the sample is less in quantity, it is amplified by polymerase chain reaction (PCR).

DNA Analysis

a. Early Markers

Discovery of ABO blood groups was the first step towards identification of individuals with limited polymorphism [17] but the polymorphism was very limited in ABO blood grouping. The advent of DNA- based markers had given unlimited extent of polymorphisms sufficient to differentiate one individual from another.

b. Southern hybridization based DNA profiling

Eukaryotic genomes are full of regions of repeat DNA sequences, often referred to as satellite DNA. The core repeat unit for a medium length repeat referred to as a minisatellite or variable number of tandem repeats (VNTR), is in the range of approximately 10-100 bases in length. VNTR patterns are inherited and are unique [29]. The children will inherit the different combinations of paternal and maternal VNTRs and therefore their profile will be different from each other. DNA regions with 2-6 bp repeat units with maximum length of around 300 bases are called microsatellites, simple sequence repeats (SSRs) or short tandem repeats (STRs). These show a high degree of polymorphism. SSRs can be detected by specific dyes or by radio-labeling after separation on gel electrophoresis [14]. DNA markers are small DNA probes that will bind to a complementary sequence. A distinctive pattern for an individual is created by a series of probes bound to a DNA sample. The fact that DNA is a very stable molecule makes this technology more effective. This is equally good for archived specimens.

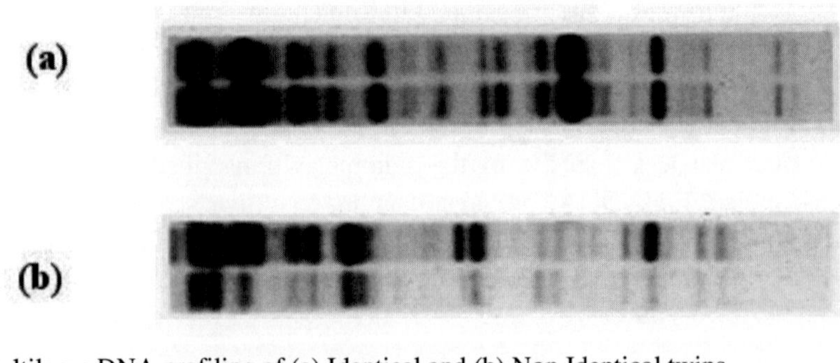

Figure 1. Multilocus DNA profiling of (a) Identical and (b) Non Identical twins.

Hybridization-based DNA fingerprinting involves use of two types of probes called multi locus probes (MLP), which identify repeating sequences in genomic DNA digested by restriction enzymes followed by Southern blot analysis. The radio labeled probes hybridize with a series of tandemly repeated sequences with a size range greater than 20 Kb. The resulting autoradiograph consists of a ladder-like series of bands. The MLP system is, however, not much successful as it is demanding with respect to quantity and quality of DNA and is more time consuming. Moreover results are not suited to computer databases. The technology really progressed after the advent of single locus probe (SLP). In the SLP system probes were isolated which identified repeat sequences in individual mini-satellites to produce a maximum of two bands per probe. Although the targeted mini-satellites showed considerable variation in their repeat numbers, a single locus was not sufficient for individualization. A number of such loci have to be analyzed.

c. PCR based DNA Profiling

DNA fingerprinting technology took a dramatic change in the early 1990's with the advent of polymerase chain reaction (PCR) technique that enabled amplification from tiny amounts of template DNA. The method was fast, reliable and efficient to work even with degraded samples. In PCR based technique the time required to analyze a sample was significantly reduced [28]. It was found to be more suited than the previously used MLP and SLP systems. Many polymorphic loci like D1S80, HLADQα were used for analysis. There were many reports claiming that it was not much useful for forensic purposes [25]. In 1992, the use of a new compatible form of genetic profiling called short tandem repeats (STR) was introduced. Due to the small size they were extremely stable and could be detected even with old and degraded samples. By 1998 STR analysis became an established feature which basically involved study of panel of polymorphic loci THO1, VWA, FGA, D21S11, D8S1179 and D18S51 now further increased to D7S820, D13S317, D5S818, CSF1P0, D16S539, D2S1338, D19S433, TPOX with Penta E and Penta D [11]. The recent new innovation in DNA chip technology (in which thousands of short DNA sequences are embedded in a tiny silicon chip) enables much more rapid, cheaper analysis using many more probes [19]. A flowchart is discussed in Figure 2 which shows the steps involved in DNA profiling.

EXISTING DNA TECHNIQUES

Restriction fragment length polymorphism (RFLP) is a technique for analyzing the variable lengths of DNA fragments that result from digesting a DNA sample with special restriction endonucleases. Amplification fragment length polymorphism (AFLP) is a PCR-based derivative method of RFLP in which sequences are selectively amplified using primers. It is a reliable and efficient method of detecting molecular markers. Random amplified polymorphic DNA (RAPD) is one of the most commonly used primary assays for screening the differences in DNA sequences of two species of plants. It involves random amplification [9]. HLA-DQα analysis was also used for identification for quite long time [23]. These techniques became obsolete and were replaced by various techniques like microsatellite analysis and single nucleotide polymorphisms etc. Some are listed below.

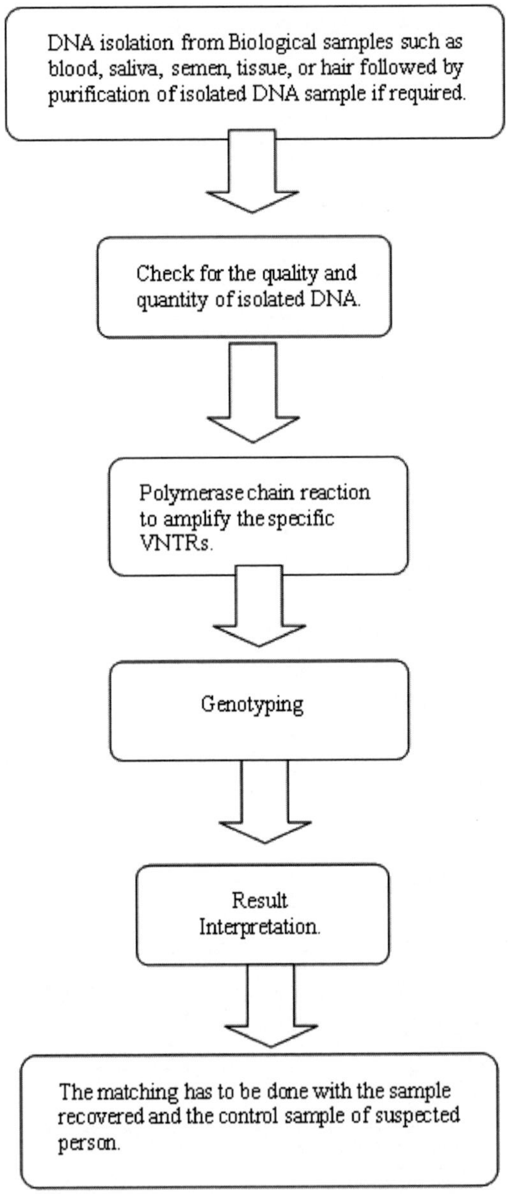

Figure 2. Flowchart of DNA profiling.

a. Capillary Electrophoresis

Capillary electrophoresis describes a technique used to separate a variety of samples. These analyses are performed in narrow tubes and can result in the rapid separation of different compounds. DNA typing with short tandem repeat (STR) markers is now widely used for a variety of applications.

b. Short Tandem Repeat Analysis

STRs have discrete alleles. Variability in STR regions can be used to distinguish one DNA profile from another [11]. The chance that two individuals will have the same 13-loci DNA profile is one in one billion. Microsatellite typing can be multiplexed and semi-automated. Sophisticated softwares have been developed for STR analysis [32]. This is done in two steps by two different software programs. Genescan software (Applied Biosystems) is used to size the DNA fragments. The resulting pherograms are then imported into the second software program, Genotyper. This program determines each sample genotype by comparing the sizes of alleles observed in a standard allelic ladder. Genescan program is used to compare the individual's profile (Figure 3). In genotyping the alleles are denoted by peaks (Figure 4). The preferred system, for STR genetic typing, was primers with fluorescent dye and automated DNA sequencers. Multiplexes are analyzed and typed using automated sequencing equipment. STR base provides complete information about different STR markers available [7].

1: DNA Profile of the mother
2. DNA Profile of the child
3. DNA Profile of the father

Figure 3. Genescan analysis of STR markers in a Paternity case. Each band present in child's lane is comparable with either of the parents.

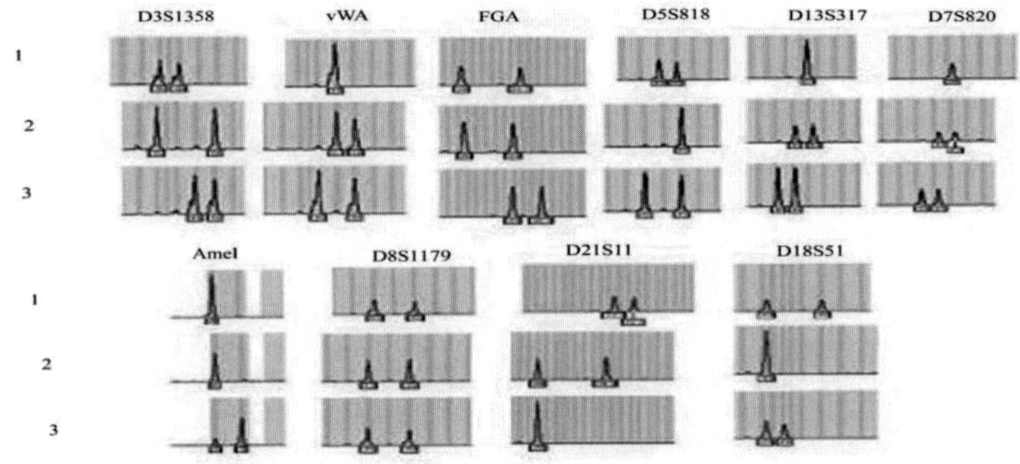

1. Genotype of the mother
2. Genotype of the child
3. Genotype of the father

Figure 4. Electropherogram of genotype of STR markers in samples of a paternity case where child is the biological offspring of the parents. The names of analysed loci are mentioned in the figure e.g. D3S1358. **Each peak present in child's profile is comparable with either of the parents.**

Other Methods include the use of single nucleotide polymorphs (SNPs), DNA amplification fingerprinting (DAF) and their offshoots. DAF involves amplification of genomic DNA which is directed by only one oligonucleotide primer of arbitrary sequence [8].

c. Mitochondrial DNA Analysis

Mitochondrial DNA is inherited **maternally and doesn't segregate indepe** y. Therefore it gives excellent record of matrilineage. Degraded samples quite oft :k nucleated cellular material, and cannot be analyzed with STR and RFLP. Mitochondrial DNA analysis can be used to examine such DNA samples as it has high copy number and has no recombinational events. Occurrence of heteroplasmy is the special feature of mitochondrial-DNA . It can be used as an important tool for identification, matrilineage and evolutionary studies. Mitochondrial DNA can also help in studying variety of human diseases, associated with mitochondrial disorders [27].

d. Y-Chromosome Analysis

Exclusively present in men, the Y chromosome is passed directly from father to son. Analysis of genetic markers on the Y chromosome is especially useful for tracing relationships among males or for analyzing biological evidence as e.g. rape cases.

e. Single Nucleotide Polymorphism Typing

Single nucleotide polymorphism (SNP) is DNA sequence variations that occur when a single nucleotide (A, T, C, or G) in the genome is changed and this can be used as marker. SNPs have lower heterozygosity as compared to STR. SNP is binary marker. Forensic SNP typing provides a lot of help to forensic scientists in mass disaster tragedy [3]. For identifying SNPs many polymorphic sites have to be **analyzed.** Mutation analysis is one of the major thrusts in functional genomics and microarrays have been widely used for this, which helps in monitoring the whole genome on a single chip and in studying the interactions of thousands of genes simultaneously. In literature various terminologies have been used to describe this technology as DNA chip, DNA microarray etc [19].

Forensic studies are mainly done by commercially developed autosomal STR multiplexes, autosomal SNPs and markers on Y chromosome and mitochondrial DNA [20].

ESTABLISHMENT OF DATABANKS

DNA intelligence databases are modern tools in investigating crime [4-5, 12, 21, 22, 24, 30]. The first of the national DNA databases involving STR data was formed in the United Kingdom in 1995. In the United States, the Federal Bureau of Investigation (FBI) has created a national database of genetic information called the National DNA index system. This database contains DNA obtained from convicted criminals and from evidence found at crime scenes. The Combined DNA index system (CODIS) utilizes computer and DNA technologies together. All DNA profiles stored in CODIS are generated using STR (short tandem repeat) analysis. The current version of CODIS uses two indexes where biological evidence is recovered from the crime scene. The convicted offender index contains DNA profiles of individuals convicted of sex offenses and other violent crimes. The forensic index contains DNA profiles developed from crime scene evidence. A computer software will search these two indexes for matching DNA profiles.

Applications of DNA Fingerprinting to Plants and Animals

DNA typing in plants and animals is a vast topic. Plant DNA fingerprinting is useful in obtaining molecular markers to identify cultivars. This also can be helpful in providing evidence in crime cases. RAPD can be a better choice for identification of plant strains. DNA profiling of plants can be used in solving civil disputes over the identity of commercially important cultivars [18].

The genotypic information is used for identification of animals [1], forensic analysis [31] and for tracing specific animals to finished meat products [26], pedigree analysis, selection and breeding programs. Forensic analysis can be of immense help for poaching cases and to check the illegal trade of endangered species. Cytochrome b region has been used for identification of the endangered animal species [2].

Ethical, Legal, and Social Concerns about DNA Fingerprinting

DNA fingerprinting can provide insights into many aspects of a person and their families including susceptibility to particular genetic disorders, legitimacy of birth, and fertility. This increases the potential for genetic discrimination. DNA typing should be carried out meeting all the international standards and following all the ethical, legal and social concerns of DNA typing. Data banking also involves some ethical issues and requires a balance between human rights and justice.

Some Examples of Usage of DNA Fingerprinting in Forensic Identification

The DNA fingerprinting technique is widely used to identify criminals, to exonerate persons wrongly accused of crimes, to identify crime and catastrophe victims in mass disasters, to establish paternity and other family relationships, to identify endangered species and wild life poaching cases, to match organ donors with recipients in organ transplantation, to determine pedigree for seed or livestock breeds, to identify different strains and cultivars to protect the rights of plant breeders and to provide personal identification.

Some Important Case Studies Where DNA Fingerprinting Played a Major Role

Identification of the Remains of the last Russian Czar

Nicholas II, the last Czar of Russia and rest of his family members were killed in Yekaterinburg during revolution. DNA tests done much later suggested that the skeletal remains found belong to two children of Nicholas II.

O.J. Simpson Case in Los Angeles

This case has been described as the most publicized criminal trial in history, in which O. J. Simpson, former American football star and actor, was brought to trial for the murder of his ex-wife and her friend.

Clinton-Lewinsky Case

The Lewinsky scandal was a political sex scandal emerging from a sexual relationship between United States President Bill Clinton and a 22-year-old White House intern, Monica Lewinsky. The news of this extra-marital affair and the resulting investigation based on DNA testing eventually led to the truth.

Thomas Jefferson-Sally Hemings Case

Jefferson DNA data was tested in 1998 in an attempt to end the long controversy regarding whether or not United States President Thomas Jefferson fathered children by his slave Sally Hemings. Researchers tested Y-chromosomal DNA from living male claimed-descendants of Jefferson and Hemings. The study concluded that Jefferson was the probable biological father of Eston Hemings.

Identification of the Bodies in World Trade Center Attack

Identification of the nearly 3000 victims of the World Trade Center attack, represented by about 15,000 body parts, rested heavily on successfully performed DNA testing.

Assassination Case of Rajeev Gandhi, Prime Minister of India

Independent India's ninth Prime Minister Rajiv Gandhi was assassinated by Thenmuli Rajaratnam alias Dhanu, a suicide bomber of the Liberation Tigers of Tamil Eelam (LTTE) at Siriperumbudur near Chennai. Her body was identified using this technology.

LIMITATIONS OF DNA TECHNOLOGY

DNA fingerprinting is sometimes controversial. One is the accuracy of the result, always in the form of some probability value. Possibilities of misuse of the technique, chances of manual error and contamination are likely. DNA fingerprinting is often performed in private laboratories that may not follow uniform testing standards and quality controls. Human error could lead to false results. DNA fingerprinting is expensive. Awareness is still lacking about this technology among law enforcing agencies and judiciaries. The probability of a DNA fingerprint belonging to a specific person needs to be reasonably high especially in criminal cases. High probability can be achieved by increasing the number of polymorphic loci analyzed. Population structure can cause variation in allele frequencies between subpopulations [15]. Some technical difficulties like errors in the hybridization and probing process will also affect the probability. Population studies should be carried out and reproducibility of the results should be confirmed. DNA testing faces difficulties of impurities and PCR inhibitors present in the sample. Mixed samples can be challenging to detect and interpret. If mutation occurs at the primer binding site it can result in its amplification. This condition is referred as null allele or allele drop-out. Sometimes stutter bands, occur because of strand slippage, are peaks several bases smaller than STR allele peak. These bands effect the interpretation of DNA profiles especially in mixed samples (John M. Butler-Forensic DNA typing) [14]. Limited resources and poor infrastructure of different laboratories **affect** the efficiency of the DNA typing laboratories.

DNA fingerprinting is reliable technique for identification not withstanding these limitations. There are many international organizations involved in quality control and uniformity of DNA fingerprinting. These are technical working group on DNA analysis methods (TWGDAM), The DNA advisory board (DAB), national institute of standards and technology (NIST) in USA, international society for forensic genetics (ISFG) in Europe.

Conclusion

DNA profiling is the technique of choice today for identification if used intelligently and efficiently. The need is to spread the awareness about the technique among common man.

Acknowledgment

This article is a modified version of the review article "DNA Fingerprinting in the Criminal Justice System: An overview" published in DNA and Cell Biology vol. 25, Number 3, 2006 and is reprinted with permission from Mary Ann Liebert, Inc. Publishers. My special thanks to them. Draft of this manuscript was significantly improved through valuable consultation with Dr T Ramasarma.

References

[1] Barallon, R. Species determination by analysis of the Cytochrome b gene in Lincoln, P. J. and Thompson, J. (Eds.), Forensic DNA Profiling Protocols, (1998). Humana Press, Totowa, NJ.

[2] Branicki.W, Kupiec,T & Pawlowski,R Validation of cytochrome b sequence analysis as a method of species identification. *J. Forensic Sci.* (2003), 48, 83-87.

[3] Bruce Budowle, Frederick R. Bieber and Arthur J. Eisenberg, Forensic aspects of mass disasters: Strategic considerations for DNA-based human identification. *Legal Medicine*, July 2005, 7, Issue 4 pp 230-243.

[4] Budowle, B and Moretti,T.R. Proceedings of the Ninth International Symposium on Human Identification. Madison, Wl: Promega Corporation, (1998), 64-73.

[5] Budowle, B, Moretti,T.R, Niezgoda, S.J. and Brown,B.L. Proceedings of the second european symposium on human identification. Madison, Wl: Promega Corporation,1998,73-88.

[6] Butler John M, Buel E, Crivellente F, McCord Bruce R, Forensic DNA typing by capillary electrophoresis using the ABI Prism 310 and 3100 genetic analyzers for STR analysis. *Electrophoresis,* 2004, (25) 1397-1412.

[7] Butler, J.M., Forensic DNA Typing: Biology and technology behind STR markers. 2001, Academic Press.

[8] Caetano-Anolles G, Bassam BJ, Gresshoff PM, DNA amplification fingerprinting using very short arbitrary oligonucleotide primers. *Biotechnology,* (NY) 1991, June 9(6),553-7.

[9] Cooper, ML, RAPD analysis of southern brown bandicoots (Isodon obesulus) populations in West Australia reveals genetic differentiation elated to environmental variables 2000. *Mol Ecol.*; 9:469-479.

[10] Edwards A., Civitello A., Hammond H.A., Caskey C.T., DNA typing & genetic mapping with trimeric & tetremeric tandem repeats, *Am.J.Hum Genet.* 1991, 49, 746 – 756.

[11] Hoyle R, Forensics. The FBI's national DNA database, *Nat. Biotechnol.* 1998, (16), 987.

[12] Jeffreys, A. J., Wilson.V, and Thein. S.L., Individual specific "fingerprints" of human DNA, *Nature*, 1985 (316) 76-79.

[13] John M. Butler. Forensic DNA Typing: Biology and Technology behind STR markers 2001.(Academic Press).

[14] Lander, E S & Budowle,B. DNA Fingerprinting dispute laid to rest: *Nature*, 1994 (371) 735-738.

[15] Landsteiner.K. Uber agglutinationserscheinungen normalen menschlichen blutes. Wien Klin Wschr (1901); 14: 1132-1134.

[16] Lekha D Kumar , Kathirvel *M*, Rao *G V*, Nagaraju *J*. DNA Profiling of disputed chilli samples (Capsicus annum) using ISSR-PCR and FISSR-PCR marker assays: *For. Sci. Intl.* 2001 (116),63-68.

[17] Linacre.A, Graham.D , *Expert Rev Mol Diagn.* 2002, Jul;2(4):346-53.

[18] Mark A.Jobling and Peter Gill, Encoded evidence:DNA in Forensic analysis. *Nature,* Reviews Oct 2004 (5), 739-751.

[19] Parsoh.W, Steinlechner M, Scheithauer R, Schneider D.M. National DNA intelligence database in Europe – report on the current situation in: Proc 9th Symp Human identification, Promega Corporation,1998, 52 – 54

[20] Peeren boom.E. Central DNA database created in Germany, *Nat. Biotechnol.*, 1998, 16, 510 – 511.

[21] Saiki R.K, Bugawan T.L, Horn G.T, Mullis K.B, Erlich H.A, Analysis of enzymetically amplified α-globin and HLA-DQα DNA with allele – specific oligonucleotide probes, *Nature*, 1986, (324), 163 – 166.

[22] Schneider P.M, DNA databases for offender identification in Europe-the need for technical, legal & political harmonization in: Proc Europ 2nd Symp Human identification, Promega Corporation, 1998, 40 – 44

[23] Varsha et al, D1S80 variation in random Indian populations, *Proceedings of X Intl. Symposium of Human Identification*, FL, USA 1999.

[24] Varsha et al, Identification of cooked meat sample by 12SrRNA sequence analysis. Proceedings of 55th annual meeting of American Academy of Forensic Sciences, 2003, chicago,USA.

[25] Wallace, D. C,Mitochondrial diseases in man and mouse, *Science*, 1999. 283:1482-1488.

[26] Weber J.L, May P.E, Abundant class of human DNA polymorphisms which can be typed using polymerase chain reaction, *Am.J.Hum.Genet.* 1989, 44 388 –396.

[27] Weising.K, Nymbom.H, Wolff, K and Mayer,W, DNA Fingerprinting in plants and fungi. 1995, CRL Press, Boco Raton.

[28] Werret D.J, The National DNA database, *Forensic Sci. Int.* 1997,88, 33-42

[29] Zehner, R., Zimmerman S. and Mebs. D, RFLP and sequence analysis of the cytochrome b gene of selected animals and man: methodology and forensic application, *Int. J. Leg. Med.*, 1998, 111:323-327.

[30] Ziegle, J.S., Su,Y.,Corcoran, K.P., Nie,L., Mayrand, P.E., Hoff,L.B., McBride,L.J., Kronick, M.N.and Diehl,S.R., *Genomics*, 1992, 14,1026-1031.

In: Recent Trends in Biotechnology, Volume 1
Editors: M.P. Singh, A. Agrawal and B. Sharma

ISBN 978-1-60876-148-7

Chapter 6

MUSHROOM BIOTECHNOLOGY

M. P. Singh*, Abhishek K. Srivastava, Sanjay K. Vishwakarma, Vinay K. Singh and Vivek K. Pandey

Department of Biotechnology, VBS Purvanchal University,
Jaunpur – 222001, India

ABSTRACT

Mushroom biotechnology is concerned with all aspects of the application of mushroom i.e. domestication of wild variety, enhanced production of nutriceuticals, bioconversion of lignocellulosic wastes into useful products, biodegradation of recalcitrant substances and bioremediation of xenobiotics. Molecular tools are being used for the improvement of mushroom strains which in turn can act as an effective weapon against malnutrition, many diseases and environmental pollution. Gene transfer using protoplast has been developed to break down the barrier of gene imposed by conventional breeding systems. Protoplast technology provides a highly feasible technique for interspecific and intergeneric hybridization. The versatile enzymatic systems of some of the mushrooms, which attack lignin and have low substrate specificity, are exploited to degrade many toxic and recalcitrant pollutants.

INTRODUCTION

Mushrooms have been object of fascination to human beings since time immemorial because of their sudden appearance along with rapidity of growth. They have intrigued us for centuries, have become part of our culture, delight us with their flavours, aromas, and therapeutic properties and also some of them frighten us due to its poisonous nature because of the presence of toxins. Presently mushrooms are among the most popular, non-conventional food accepted the world over and the increased consumer demand over the years has led to quantum jump in its production. Mushroom is a group of fungi endowed with

* E-mail: mpsingh.16@gmail.com ; mpsingh_07@yahoo.com

ability to convert inedible organic wastes into palatable food. Because of its pleasant flavour and protein it is a source of human food which can satisfy the basic human requirement for better nutrition. The therapeutic value of mushrooms is relatively untapped. Besides, the substrates on which mushroom grow are lignocellulosic wastes occur in agricultural, forestry, fruit and vegetable wastes and are found in plenty. The role of mushroom in bioconversion, biodegradation and in bioremediation has increasingly being realized in recent years. They are proving as an effective and important tool for replenishment of earth's overburdened ecosphere.

Mushroom biotechnology offer ample opportunities to convert these wastes as a vast renewable resource to meet the growing demand of food, fodder, fertilizer and energy and increase the efforts to recycle them. It is also consistent with the emerging view that ecologically oriented society must use its resources rather than discarding them as useless materials.

MUSHROOM AND MUSHROOM BIOTECHNOLOGY

The term 'mushroom' has been perceived differently by different people in the world and consequently defined accordingly. According to Atkinson [1], "the mushroom belongs to Basidiomycetes, and toadstool is regarded as a synonymous term, since, there is strictly speaking, no distinction between a mushroom and a toadstool". However, Snell and Dick said, "A mushroom may be edible, poisonous, unpalatable, tough etc., but popular usage applies the term to edible ones, calling others 'toadstool'[19]. Chang and Miles defined, "the mushroom is a macrofungus with a distinctive fruiting body which can be either epigeous or hypogeous and large enough to be seen with the naked eye and to be picked by hand" [3]. Chang further, divided mushrooms into four categories: (1) those which are fleshy and edible fall into the edible mushroom category, e.g., *Agaricus bisporus*; (2) mushrooms which are considered to have medicinal applications, are referred to as medicinal mushrooms, e.g., *Ganoderma lucidum*; (3) those which are proven to be, or suspected of being poisonous are named as poisonous mushrooms, e.g., *Amanita phalloides*; (4) a miscellaneous category which includes a large number of mushroom whose properties remain less well defined. These may tentatively be grouped together as 'other mushrooms' [3].

Botanically mushrooms are fungi which lack chlorophyll, hence, unable to synthesize food using solar energy. Mostly they are saprophytic which derive their energy from organic substrates on which they grow. Some are parasitic. The number of recognized mushroom species has been reported to be 14000 [7]. Currently there are still many unknown species yet to be discovered (about 126,000) [4]. According to Benzamin [2], over 2000 species are probably edible and 400 species are poisonous. Out of the over 2000 edible species, 100 are widely picked, 15-30 species are commonly eaten, 80 species are experimentally grown, 40 species are commercially cultivated and 5-6 species are produced on mass scale. Some of the important mushrooms cultivated in India as well as other parts of the world are given in table-1.

Table 1. Some common cultivated mushrooms

Common Name	Scientific Name
Button/European/Temperate mushroom	*Agaricus bisporus*
Button/Edulis/hot weather mushroom	*A. bitorquis*
Oyster mushroom	*Pleurotus spp.*
Shitake mushroom	*Lentinus edodes*
Black ear	*Auricularia polytricha*
Paddy straw/Chinese/tropical	*Volvariella volvacea, V. diplasia*
White milky mushroom	*Colocybe indica*
Winter mushroom	*Flammulina vetutipes*
Brown cap/giant mushroom	*Stropharia rugoso-annulate*
Nameko mushroom	*Pholiota nameko*

Based on ignorance, the common myths and misconceptions about the mushroom i.e. most mushrooms are poisonous, deadly mushrooms are instantly lethal, it is dangerous to touch a poisonous mushroom etc. were dispelled with understanding in mushroom biology and advancement in technology and advent of mushroom biotechnology. Mushroom biotechnology is concerned with application of molecular tools in the improvement of mushrooms and subsequently the use of mushroom in combating malnutrition, get rid of many diseases, biodegradation and bioconversion of lignocellulosic wastes into food, fodder and energy and bioremediation of xenobiotics.

MUSHROOM NUTRICEUTICALS

The term 'mushroom nutriceuticals' has been coined to embody both the nutritional and medicinal features of these biological response modifiers extractable from either the mycelium (culture fluids) or fruit bodies of the mushroom [5]. Currently, mushrooms represent one of the most rapidly growing sources of extractable both for inclusion in food supplements to enhance health and fitness, and in the prevention and treatment of human disease conditions [4].

Generally, edible mushrooms possess all the three properties of food – nutrition, taste and physiological functions. They are rich in protein, chitin, vitamins and minerals; low in fat with high proportions of unsaturated fatty acids, and calories; and contain no cholesterols. Mushrooms serve as delicious foodstuff and also flavouring substances because of their unique flavours. Besides the volatile eight-carbon compounds, the typical mushroom flavour consists of water-soluble taste components such as soluble sugars, polyols, organic acids, free amino acids and 5-nucleotides. They contain several physiologically active substances including high molecular weight polysaccharides (mainly β-D-glucans), heteroglucans, chitinous substances, peptidoglucans, proteoglucans, lectins, RNA components, dietary fibre; and low molecular weight organic substances, such as terpenoids, steroids, novel phenols [4].

Some of the mushrooms have shown to improve the general health of AIDS patients. *Ganoderma* species contain biocompounds and micronutrients (e.g. selenium), which boost the body's immune systems. In AIDS patients, the intestinal lining gets damaged, making it difficult for food to be absorbed into the body and causing many patients to die simply from starvation. Patients given *Ganoderma* products were found to have their damaged lining of the intestine repaired, improving food absorption, and consequently the health of the patients. For example, patients' skin complexion improved, warts disappeared and mouth sores healed [4].

PRODUCTION TECHNOLOGY

The consumption of mushroom by human beings from nature is probably as old as human being itself. Due to its pleasant flavour and taste people tried to domesticate mushroom in latter stage. The earliest cultivated edible fungus was a species of *Auricularia* which was cultivated about 1000 years ago in China. The next species to be cultivated was *Lentinus edodes*. This occurs almost 900 years ago, also in China [6]. *Agaricus bisporus* which is leader in the global mushroom industry today in terms of its production was first cultivated in France in 1650. Since then a number of other mushrooms have been domesticated. Mushrooms are cultivated on lignocellulosic substrates. Some of the mushrooms can be cultivated in composted (fermented and partially degraded) substrates primarily cereal straws and other supplements specially the button mushroom (*Agaricus* spp.). Use of composted substrate is mainly aimed at facilitating the growth of mushroom mycelium to the practical exclusion of microorganisms. However, other mushrooms can be grown on fresh, lignocellulosic substrates i.e. *Pleurotus, Lentinus, Flammulina, Auricularia, Pholiota, Tremella, Calocybe, Ganoderma, Coprinus* etc.

Over 90% of total world production of cultivated edible mushroom comes from 6 genera - *Agaricus, Lentinus, Volvariella, Flammulina, Auricularia* and *Pleurotus.* Revolution in biotechnology opened up new vista in mushroom science for the conservation of germplasm, development of high yielding variety with yield stability, high biological efficiency having other desirable traits like rich in nutrients and taste, strong aroma and longer shelf life.

The application of molecular biology techniques to the genetic improvement programmes has introduced innovative DNA based markers and provides much needed fillip to the ongoing developments in genetics and breeding of cultivated mushroom [20]. Molecular markers have heritable differences in nucleotide sequences of DNA from two individuals, which follow a simple Mendelian pattern of inheritance. These DNA markers are detected employing two basic techniques of southern blot hybridization and polymerase chain reaction (PCR). The application of molecular markers has hastened the development of genetically improved strains and hybrids. These DNA based markers are used in the assessment of genetic diversity and selection of parents, identification of homokaryons and confirmation of hybridization, genomic relationship, constructions of linkage map, breeding behavior and sexuality, DNA fingerprinting and mushroom variety protection. Molecular markers such as allozymes, RFLPs, RAPDs, repetitive DNA sequences, ITS and AFLP offers several advantages over other markers. These include - abundance, codominance, phenotypic neutrality, absence of epistasis and developmental stage, tissue and environment independent expression. Large number of molecular markers can be obtained by using DNA probes from a variety of sources in combination with any of the commercially available enzymes and various DNA primers [20].

PROTOPLAST TECHNOLOGIES

The usefulness of the protoplast technology in several areas of mushroom biology has been well established. The major area of interest has focused on its application as a tool for genetic manipulation. Protoplast technology is non-conventional method of genetic manipulation and play important role in genetics and breeding of mushrooms. Gene transfer using protoplast has been developed to break down the barrier of gene imposed by conventional breeding systems. The development of a gene transfer system in edible fungi would be useful in genetic recombination and strain improvement [21]. In the Basidiomycetes the most exciting aspect of this endeavour concerns the generation of novel interspecies hybrids [12]. It is important to note that strain improvement by protoplast fusion have been performed in closely related species. Successful recombination between distantly related species provides the opportunity for a broadening of the gene pool. The uptake of isolated cell organelles by protoplasts may allow the exploitation of the protoplast fusion technique on a more refined level. DNA- mediated transformation related to gene vectors advances our understanding of the basic molecular biology of higher fungi. Development of this method also opens up many possibilities for the commercial production of medically important substances found in mushrooms [21].

Protoplast technology provides a highly feasible technique for interspecific and intergeneric hybridization among the economically important mushroom. Hybridization using protoplast fusion has significance for breeding between intraspecific incompatible Basidiomycetes species, feasibility of interspecific and intergeneric hybridization, development of multifunction such as the formation of physiological active substances (medicinal compounds), aromatic and growth acceleration, growth acceleration of mycorrhizal mushrooms and genetic analysis [11].

LIGNOCELLULOSE DEGRADATION THROUGH MUSHROOM

Lignocelluloses comprising cellulose, hemicellulose and lignin are the most abundant renewable organic compounds available in the biosphere, being produced through photosynthesis and allied biological processes [10,13,14]. Significant development in science and technology has led to quantum jump in the agricultural commodities production. This in turn has further accelerated the lignocellulosic biomass accumulation. Major part of these are burnt resulting multifaceted hazards including oxygen deficient environment, respiratory diseases (such as allergic bronchitis, asthma, tuberculosis) and poor visibility in night. Its indiscreet burning increases carbon dioxide and unburnt carbon particles in air living silicon rich ash in the fields, degrading top soil resulting in both poor utilization of lignocelluloses as well as wastage of energy. Mushroom growing on lignocelluloses provides nutritional rich food, fodder, fertilizer and energy thereby recycling into useful products. Figure 1 illustrates the biodegradation of lignocelluloses into useful products.

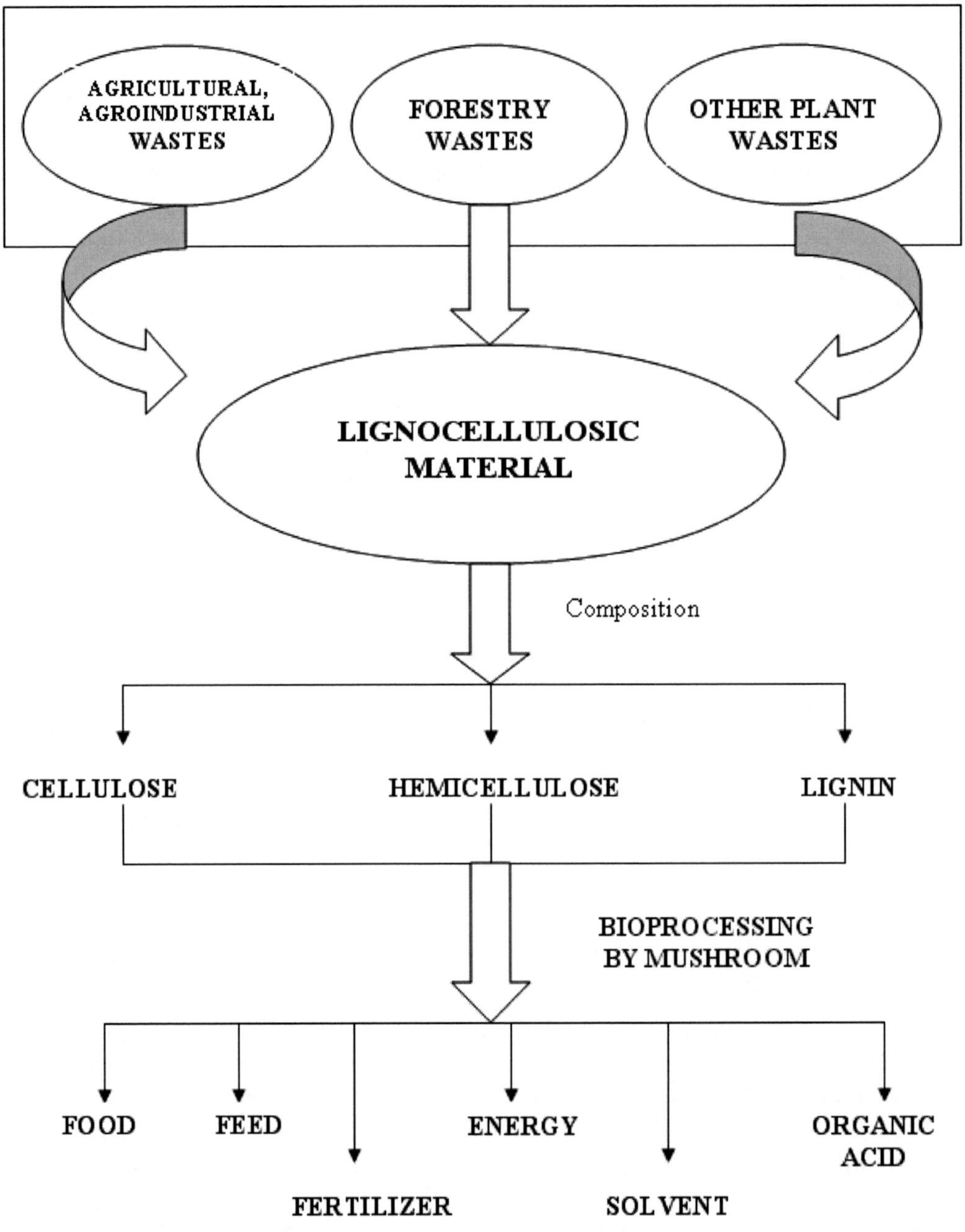

Figure 1. Bioprocessing of lignocelluloses by mushroom.

Exctracellular enzymes secreted by mushrooms play a key role in the bioconversion and biodegradation of lignocellulosic wastes into useful products [15,16]. These enzymes are cellulases, xylanases and ligninases. Cellulases or cellulolytic enzymes minimally composed of endoglucanase, exoglucanase and β-glucosidase are involved in biodegradation of cellulose. These enzymes act synergistically and hydrolyse the β-1 and 4 glucose linkages. Cellulose is randomly cleaved by an endoglucanase to produce free chain ends. An exoglucanase cleave cellobiose units from the nonreducing ends. The soluble disaccharide is then hydrolyzed by β-glucosidase. Xylanases enzymes are responsible for the hydrolysis of hemicellulose component of lignocellulose. The enzyme hydrolyses branched xylan and reduces its chain length. Ligninases or lignolytic enzymes degrade lignin to humus, water and carbon dioxide. The main enzymes responsible for depolymerisation of lignin are laccase, manganese peroxidase (MnP) and lignin peroxidase (LiP). In addition to these key enzymes, veratryl alcohol (VA), a product lignin degradation and glyoxal oxidase (GLox), an extracellular peroxide-generating enzyme, also have important role in lignin degradation.

Biodegradation of a wide range of lignocellulosic waste through mushroom cultivation is significant as it not only leads to formation of simpler compounds but protein rich food. During cultivation of oyster mushroom on various lignocelluloses, cellulose, hemicellulose lignin and carbon content of substrate decrease, while nitrogen and ash content increase and C/N ratio come closer. The progressive breakdown of cellulose, hemicellulose and lignin is correlated with apparent increase in yield and bioefficiency of oyster mushroom [13,14]. During biodegradation, about 50% of the substrate is liberated to CO_2, 20% as water and 10% as fruit bodies (Figure 2).

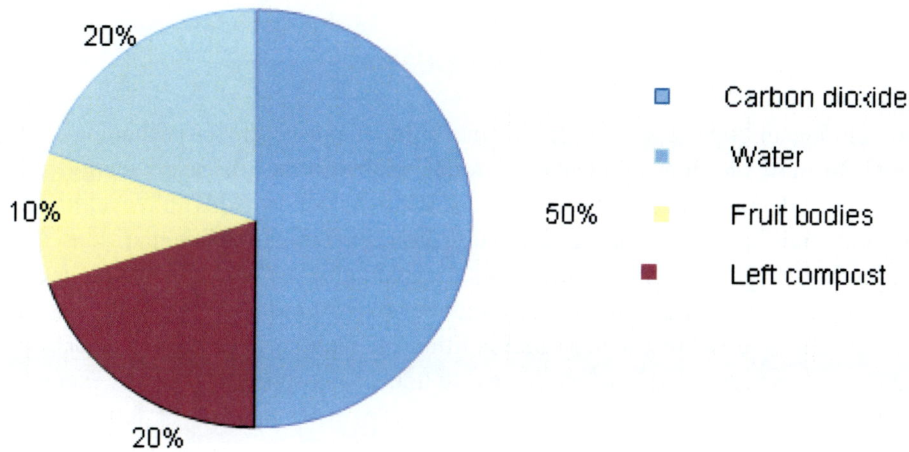

Figure 2. Biodegradation of lignocelluloses.

Bioremediation through Mushroom

Bioremediation is defined as the use of living microorganisms to reduce or eliminate environmental hazards resulting from accumulation of toxic chemicals and other hazardous wastes. In other words it is the application of biological processes to the treatment of pollutants. Most research within the field of bioremediation has focused on bacteria, but recently, fungal bioremediation (mycoremediation) is attracting the interest of scientists and environmentalist all over the world. All the three types of fungi i.e. saprophytic, parasitic and mycorrhizal are used in bioremediation processes. Among them white rot fungi in general and oyster mushroom (*Pleurotus* spp.) in particular possess the versatile enzymatic system [17] which degrades pollutants effectively and efficiently. Four main genera of white rot fungi which have shown potential for bioremediation are *Phanerochaete, Trametes, Bjerandera* and

Pleurotus [8]. These fungi can not use lignin as a source of energy, however, and instead require substrates such as cellulose or other carbon sources. Thus, carbon sources such as corncobs, straw, sawdust or any lignocellulosic residues can be easily used to enhance degradation rates by these organisms at polluted sites. The main mechanism of biodegradation employed by this group of fungi, however, is the lignin degradation system of enzymes. These extracellular lignin modifying enzymes (LMEs) have very low substrate specificity which makes them suitable for degradation of many toxic, hazardous compounds like polycyclic aromatic hydrocarbons (PAH), polychlorinated biphenyls (PCB), trinitrotoluene (TNT), DDT etc. having structural similarities with lignin. Singh *et al*, [18] reported effective decolourization of azo dyes by nine species of *Pleurotus*. As these dyes are synthetic, they are generally toxic, mutagenic, carcinogenic and resistant to degradation. Bioremediation involves use of mushrooms to turn these pollutants into non-toxic harmless substances. Mushrooms have been shown to accumulate radionuclide metals and even rare earth elements. They are potentially useful in detecting the presence of deleterious compounds.

Conclusion

Revolution in technology has led to the birth of mushroom biotechnology which has potential to mitigate the problems of food, malnutrition, disease, energy crisis and environmental degradation. Mushroom biotechnology is concerned with recycling of organic wastes and conversion of problematic wastes into useful products like nutritious food, feed, fodder and fuel. It is also concerned with exploitation of medicinal and energy tonic properties and cosmetics as well as nutritious beverages products. Mushrooms are not only used as an effective weapon against malnutrition but also an important tool for restoration, replenishment and remediation of earth's overburdened ecosphere. This is an example of low cost, viable, multipurpose technology to meet the growing demand of food, feed, fodder, fertilizer and energy.

References

[1] Atkinson, G.F. 1961. *Mushrooms (Edibles and otherwise)*. Hafner publishing company, New York.

[2] Benjamin, D.R.1995. Incidence of mushroom poisoning. In: *Mushrooms: Poison and Panacea*. S.W.H. Freeman and Company, New York, pp. 152-161.

[3] Chang, S.T.1993. Mushroom and mushroom biology. In: S.T. Chang and J.A. Buswell (eds.), *Genetics and Breeding of Edible Mushrooms*. Gordon and Breach Science Publishers, Switzerland, pp. 1-13.

[4] Chang, S.T. 2007. Development of the world mushroom industry and its role in human health. In: *International conference on mushroom biology and biotechnology, Sovenir-cum-abstracts*, NRCM, Solan, India, pp. 3-16.

[5] Chang, S.T. and Buswell, J.A. 1996. Mushroom nutriceuticles. *World J. Microbiol. Biotech.* **12**: 473-476.

[6] Chang, S.T. and Miles, P.G. 1987. Historical record of the early cultivation of *Lentinus* in China. *Mushroom Jr. of the Tropics*. **7**: 31-37.

[7] Hawksworth, D.L. 2001. Mushrooms: The extent of the unexplored potential. *Intl. J. Medicinal Mushrooms* **3**:333-337.

[8] Hestbjerg, H.P., Willumsen, A., Christensen, M., Anderen, O. and Jacobsen, C.S. 2003. Bioaugmentation of tar-contaminated soils under field condition using *Pleurotus ostreatus* refuse from commercial mushroom production. *Environmental Toxicology and Chemistry*. **22**(4): 692-98.

[9] Kaunda, W. 2005. *Ganoderma* clinical testing in Zambia: a presentation made at UNDP/UNOPS ZERI Regional Project Consultative Workshop. Windhoek, Namibia. October 24-28, 2005.

[10] Kuhad, R.C. and Singh, A. 1993. Lignocellulosic biotechnology: current and future prospects. *Crit. Rev. Biotechnol.* 13:151-172.

[11] Ogawa, K. 1993. Interspecific and intergeneric hybridization of edible mushrooms by protoplast fusion. In: S.T. Chang and J.A. Buswell (eds.), *Genetics and Breeding of Edible Mushrooms*. Gordon and Breach Science Publishers, Switzerland, pp. 193-205.

[12] Peberdy, J.F. and Fox, H.M. 1993. Protoplast technology and Edible mushrooms. In: S.T. Chang and J.A. Buswell (eds.), *Genetics and Breeding of Edible Mushrooms*. Gordon and Breach Science Publishers, Switzerland, pp. 125-155.

[13] Singh, M.P. 2000. Biodegradation of lignocellulose waste through cultivation of *Pleurotus sajor-caju*. In: Van Griensven (ed.), *Science and Cultivation of Edible Fungi*. Balkema, Rotterdam, pp 517-521.

[14] Singh, M.P. and Gautam, N.C. 2004. An overview of lignocellulose biotechnology. In: N.C. Gautam. And M.P. Singh, (eds.), *Recent Advances in Biotechnology*. Shree Publishers and Distributors, New Delhi, pp. 3-20.

[15] Singh, M.P., Srivastava, A.K., Vishwakarma, S.K., Pandey, V.K., Pandey, A.K. and Singh, S.K. 2007. Extracellular enzymatic activities by *Pleurotus* species on vegetable wastes. *Mush. Res.* **16**(2): 93-97.

[16] Singh, M.P., Srivastava, A.K., Vishwakarma, S.K., Pandey, V.K., Pandey, A.K. and Singh, S.K. 2007. Extracellular enzyme profiles by white rot fungi on lignocellulosic wastes. *Poll. Res.* **26**(3): 445-448.

[17] Singh, M.P., Pandey, V.K., Pandey, A.K., Srivastava, A.K., Vishwakarma, N.K., and Singh, V.K.2008. Production of Xylanase by white rot fungi on wheat straw. *Asian Jr. of Microbiol. Biotech. Env. Sc.* **10**(4) : 859-862.

[18] Singh, M.P., Rastogi, P.C., Srivastava, A.K. and Vishwakarma, N.K. 2008. Decolorization of azo dyes by white rot fungi *Pleurotus* species. *Poll. Res.* **27**(3): 365-369.

[19] Snell, W.H. and Dick, E.A. 1971. A glossary of Mycology. Harvard University Press, New York.

[20] Yadav, M.C. 2004. Molecular Markers in Genetics and Breeding of Mushroom. In: N.C. Gautam and M.P. Singh, (eds.), *Recent Advances in Biotechnology*. Shree Publishers and Distributors, New Delhi, pp. 87-104.

[21] Yoo, Y.B. and Cha, D.Y. 1993. Gene transfer in edible fungi using protoplasts. In: S.T. Chang and J.A. Buswell (eds.), *Genetics and Breeding of Edible Mushrooms*. Gordon and Breach Science Publishers, Switzerland, pp. 157-191.

In: Recent Trends in Biotechnology, Volume 1
Editors: M.P. Singh, A. Agrawal and B. Sharma

ISBN 978-1-60876-148-7

Chapter 7

DIFFERENTIATION TO DEDIFFERENTIATION: INDUCED PLURIPOTENT STEM CELLS

Narendra Pratap Singh and Rakesh K Mishra
Centre for Cellular and Molecular Biology
Uppal Road, Hyderabad 500007, India

ABSTRACT

Treatment of tissue degenerative diseases like neurodegenerative disease, type one diabetes mellitus, spinal cord injury, chronic liver disorder, etc. and understanding various aspects of such diseases have remained as major challenges in the biomedical sciences. Isolation of embryonic stem (ES) cells opens up the possibility to meet these chanllanges although ethical and immunological issues in using these cells pose serious restrictions. Recent progress in developing methods to efficiently dedifferentiate cells from adult body into ES cells, known as induced pluripotent stem (iPS) cells, offers a promising and viable alternative. Use of iPS cells also bypasses ethical as well as immunological issues. IPS cell therapy, therefore, is being percied as the autogenic cell source for treatment of various tissue degenerative diseases.

INTRODUCTION

Mammalian development starts with the fertilization of the ovum by sperm to make a diploid zygote. Zygote is a totipotent cell, which differentiates and develops into full organism. After fertilization, zygote divides to form morula which further differentiates into blastocyst containing outer trophectoderm and inner cell mass (ICM). ICM contains many embryonic stem cells which can differentiate into all the cells of the body and some extra-embryonic tissue, whereas trophectoderm gives rise to the outer layer of the placenta (Fig1). This suggests that ES cells are pluripotent that can differentiate into all cells of the body except trophectoderm. ES cells are some time confused to be totipotent because they can differentiate into all germ layers: ectoderm, mesoderm, and endoderm. Totipotent cells have capability to form all parts of developing embryo whereas ES cells can form all germ layers but not the trophectoderm which gives rise to outer placenta. Lack of this totipotent property of ES cells does not allow us to use them as zygote, a true totipotent cell. As ES cells differentiate into different cell types/tissues and finally a complete body, the differentiation potential of cells becomes increasingly restricted and finally they get terminally differentiated (Fig 2). In this process they also form many lineage specific stem cells, called adult stem cells. Adult stem cells are hypothesized to be present almost in all the tissues of the body. Their main role is to replace cells lost by normal turnover or to repair the damages resulting from diseases or injuries. Few adult stem cells like hematopoietic, cardiac, neural, skin, epithelial, mesenchymal, mammary and endothelial stem cells have been identified from different tissues of the adult body. These adult stem cells can give rise to limited cell types related to a tissue type and that is why they are called as multipotent, oligopotent or unipotent stem cells. Regenerative property of stem cells to form variety of tissues makes them very attractive for their use in tissue engineering for treatment of diseases like neuro-degenerative diseases, diabetes, cardiovascular, kidney, liver and lung diseases. Tremendous use of ES cells in tissue engineering and regeneration has made them potentially very useful in the human health and diseases. It is primarily for these reasons that ES cell research has been at the focus of attention in recent years.

Isolation of ES Cells

One major problem in genetic manipulation and tissue engineering from ES cells *in vitro* is that, we need them in large amount. For this reason we need to have *in vitro* undifferentiated ES cells in culture. The idea of undifferentiated cells or ES cells came initially from a striking observation in 1970 where early mouse embryos grafted into adult mice produced teratocarcinomas[1-3], malignant multi-differentiated tumors containing a significant population of undifferentiated cells. Further studies on spontaneously occurring teratocarcinomas established that the undifferentiated components, embryonal carcinoma (EC) cells could be propagated in culture using fibroblast cells as feeder[3]. Most importantly, clonally isolated and expanded EC cells retained the capacity for differentiation and may produce derivatives of all three primary germ layers. Therefore, individual EC cells are self-renewing and can be considered as pluripotent stem cells. The carcinogenic nature of these cells, however, does not allow us to use them *in vivo* as substance for disease treatment.

The real breakthrough came in 1981 when Martin J. Evans isolated the first mouse ES cell lines from the ICM of mouse blastocyst using culture conditions having fibroblast feeder layers and serum. The important feature of these *in vitro* cultured embryonic stem cells was that it retained capability to enter into embryogenesis and formed all germ layers. In this experiment, ES cells isolated from mouse with agouti coat colour were injected in blastocyst of mouse with white coat colour and then transferred into surrogate mother. In those cases where ES cells contribute to embryonic development gave birth to chimeric (mixed coat colour) mouse. In the following generation of the chimera, only those cases where ES cells participated in germ cell formation gave coloured coat mouse (Figure 1)[4]. These ES cells were further used to manipulate mouse genome by transgenesis and knockout to understand function of genes in adult mouse. This important discovery of 'principles for introducing specific gene modifications in mice by the use of embryonic stem cells' was recognized by award of Nobel price in medicine in 2007.

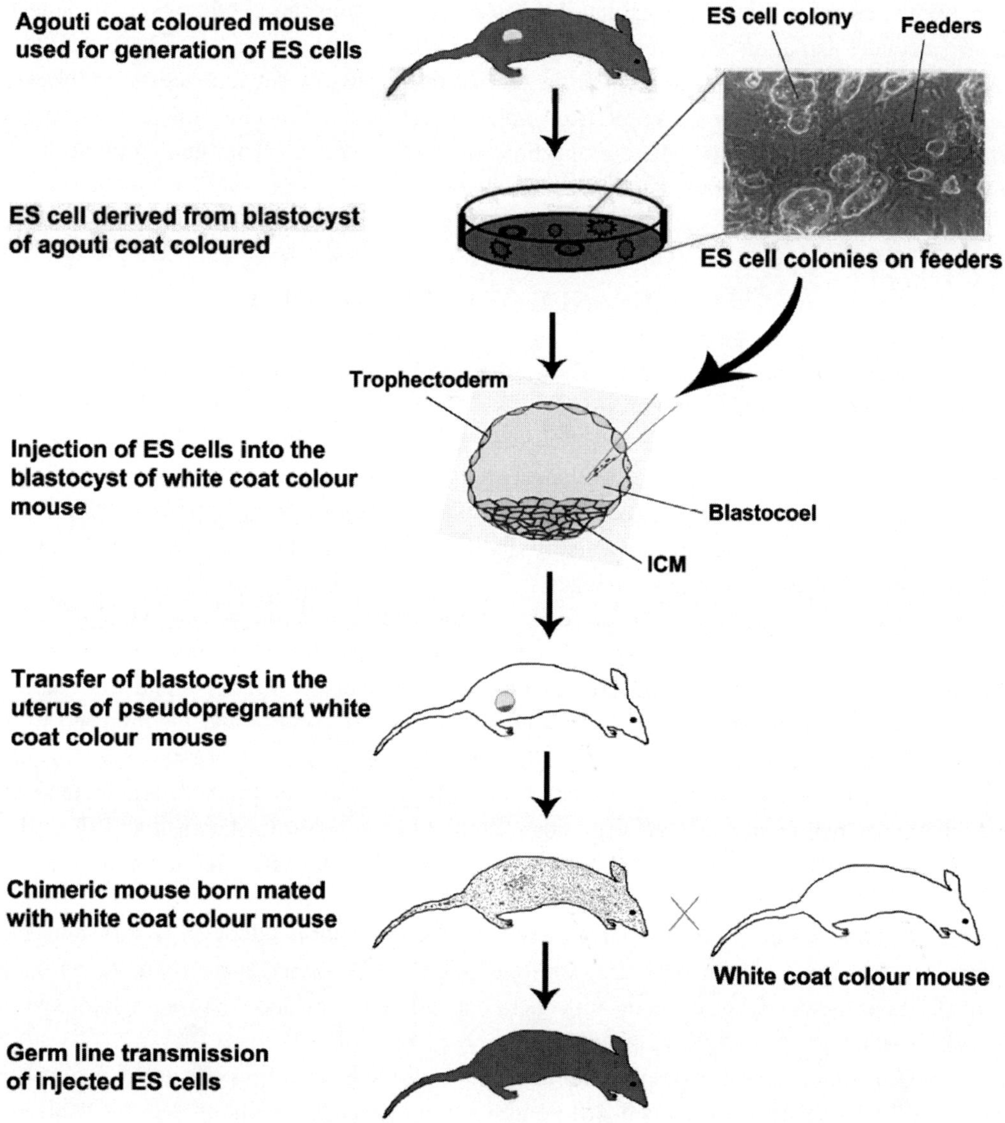

Figure 1. Use of ES cells in germ line transmission: ES cells have been isolated from blastocyst of agouti coat coloured mouse and cultured in vitro. These cells can be used for genetic alterations like transgenesis and gene knockout. For making genetically altered mouse, ES cells were injected into blastocyst of normal mouse (white coat colour) and blastocyst was transferred in uterus of surrogate mother (white coat colour). Chimeric, Go offspring of this mouse was crossed with white coat colour mouse. If germ cells of the chimeric mouse are derived from ES cells of agouti coat colour mouse they will produce offspring having agouti coat colour carrying the altered genome.

One of the major goals, in science is to help and improve human health using different tools and techniques. After the isolation of ES cells from mouse embryos scientists tried isolating ES cells from human without much success before 1998 when Thomson and colleagues isolated ES cells from human blastocyst[5]. This 17 years delay in isolating human ES cells since isolation of mouse ES cells was because of species specific differences in ES cell properties. Interestingly, ES cells from rat were isolated as recently as late 2008[6]. Initially efforts were directed to isolate human ES cells using the conditions applicable for mouse stem cell isolation without any success. Clues for culture conditions of human ES cells came from the isolation of ES cells from two primates: marmoset and rhesus monkey[7]. Human stem cells are very different from mice and so is their mechanism of self-renewal. There have been many cases where embryonic stem cells isolated from human could not be maintained in undifferentiated state in culture conditions.

MAINTENANCE OF PLURIPOTENCY

Self-renewal capacity of ES cells allows them to be cultured *in vitro*, which can be used for genetic manipulations and subsequent usage. Several features, including the presence of active telomerase protein in these cells, suggests that they can be used as immortal cells[8]. Since the discovery of ES cells, efforts are being made to understand what makes ES cells pluripotent. Understanding the molecular mechanism of pluripotency may help us to make ES cells from terminally differentiated cells by dedifferentiating them. It may also be helpful in differentiating ES cells in specific lineages.

Stability and cell division of ES cells without differentiation is achieved by use of fibroblasts as feeder cells as they provide some critical nutrients and factors for this purpose. Also medium that is "conditioned" by co-culture with various cells was found to be able to sustain ES cells in the absence of feeders. Fractionation of conditioned medium led to the identification of leukemia inhibitory factor (LIF), a cytokine that sustains ES cells undifferentiated without feeder[9]. Discovery of LIF also gave a system to understand that how ES cells maintain pluripotency. Work from different groups shows that LIF and its related cytokines act via the gp130 receptor (also known as LIF receptor) induces dimerization of gp130 receptors, which in turn regulates the Janus-associated tyrosine kinases (JAK)/the latent signal transducer and activator of transcription factor (STAT3), and ERK (extra-cellular signal regulated kinase) a mitogen-activated protein kinase (MAPK) cascades. Overall, the self-renewal of ES cells appears to depend on the balance between a high level of STAT3 and low level of ERK pathway[10-12].

While this mechanism is operational in the mouse, the human ES cells are different from mouse ES cells. In Human ES cell different signaling cascades maintain self-renewal. FGF (Fibroblast growth factor), TGF (transforming growth factor) and Activin signaling pathway molecules play major role in pluripotency of the human ES cells[13]. Basic FGF (bFGF) allows the clonal growth of human ES cells on fibroblasts in the presence of a commercially available serum[14]. At higher concentrations, bFGF allows feeder independent growth of human ES cells cultured in the same serum replacement like LIF in case of mouse ES cells. Both Activin and TGF have strong positive effects on undifferentiated proliferation of human ES cells in the presence of low or modest concentrations of FGFs[11, 15-17]. In fact all of them work together in coordinated manner to suppress BMP (Bone Morphogenetic Protein) pathway in ES cell to maintain pluripotency.

Pluripotency is a complex biological process, which is controlled by multiple factors and signaling cascades at different levels. In recent years, gene expression profile studies of ES cells have led to the identification of ES cell specific transcription factors like OCT4, NANOG, SOX2, cMyc, KLF4, and LINT 28[11, 12, 18-23]. These factors regulate various signaling cascades like Wnt, FGF, ERK, BMP and TGF, involved in the maintenance of ES cell identity[16, 17, 24]. Out of these transcription factors OCT4, SOX2 and NANOG are the most important ones as they coordinate with each other to form a tightly regulated circuit for maintaining pluripotency. They also regulate each other and in coordinated manner make large protein complexes with other factors. OCT4 and SOX2 interact with each other, whereas NANOG work with other factors to regulate the target genes involved in pluripotency. Because of almost similar binding sites, these factors regulate similar targets in the genome. They co-regulate a large set of genes involved in cellular differentiation and lineage commitment including various signaling pathways members. In summary, various pathways like Wnt, ERK, BMP, FGF, TGF, Activin, STAT3 and transcription factors OCT4, NANOG, SOX2, cMyc, KLF4, Lin28 work in coordinated manner to give a final molecular output to ES cell for maintaining pluripotency. Interestingly OCT4, SOX2 and NANOG are functionally conserved in maintenance of pluripotency in human and mouse, suggesting that the initial molecular events involved in pluripotency are similar. A concise description of different transcription factors involved in pluripotency has been given in Table 1.

UTILITY OF ES CELLS

ES cells have enormous usage in the area of basic science and medicine. *In vitro* culture of these cells enables us to manipulate the genome for understanding the basic science and then it can be applied to cure diseases. Discovery of mouse ES cells gave a different dimension to use mouse as a model to study different genes involved in human health and diseases. They have been used for studying early embryonic development, the molecular basis of pluripotency and in clarifying the function of genes by making transgenic or knockout animals. As we cannot work on human system, these cells have been proven to be very good model to understand many human specific biological processes. Similarly ES cells have also unparallel potential for use in clinics for compound screening and toxicological testing, tissue repair, and gene therapy. They have great implications in treatment of various human diseases like Parkinson's disease, type one diabetes mellitus, spinal cord injury, chronic liver disorder, etc. So far, ES cells have been successfully used to generate cardiomyocytes, neurons, endothelial cells, insulin-producing beta-Langerhans cells, hematopoietic precursors, keratinocytes, osteoblasts and hepatocytes. These initial efforts and results gave a lot of promise for medical use of these cells for treatment of several diseases. Till now, however there is no single report where ES cells have been directed to a pure lineage, The result has been a mixed population of different cell types.[25]. Several cell-signaling pathways have been found to have profound effects on lineage specific differentiation but controlling many pathways starts affecting other activity of cells including the pluripotency. It is likely that the genetic manipulation for forced expression of master transcription factors may be required for maximal efficiency. Defining all the steps require to induce any desired cell, tissue, or organ from ES cells in culture could be proposed as the ultimate challenge for the biologists.

Table 1. Key factors involve in maintenance of pluripotency

Name	Molecular signature	Function	Comments
Oct4 (octamer binding transcription factor 4)	DNA binding Pou and homeo domains	Transcriptional activator or repressor	Expressed exclusively in ICM of blastocyst. Low expression of Oct-4 in embryos fails to form the ICM and differentiate into trophectoderm. Activates fgf4 cascade to maintaining pluripotency.
Sox2 (SRY-box containing gene 2)	DNA binding HMG box domain	Transcriptional activator or repressor	Express in variety of stem cells and one of the targets of STAT3 pathway. It can activate Oct4 at blastocyst stage. Mutation in the gene lead to loss of eyes in adult and differentiation of ES cells in multiple lineages.
Nanog (named after the mythological Celtic land of the ever young, Tir nan Og)	DNA binding homeodomain	Transcriptional activator or repressor	Works in collaboration with Oct4 and Sox2 at many stem cells specific genes. Loss of Nanog function causes differentiation of embryonic stem cells into different cell types.
Klf4 (Krupple like factor 4)	Zinc finger DNA binding domain	Transcriptional activator	Integrates with Nanog transcription network and STAT3 signaling pathway to maintain pluripotency. It interacts with epigenetic modifiers to regulate the gene.
Myc (myelocytomatosis oncogene)	Helix-loop-helix DNA binding domain	Transcriptional activator or repressor	A nuclear phosphoprotein that plays a role in cell cycle progression, apoptosis and cellular transformation. Over expression leads to hyper proliferation and cancer. It can regulate STAT3 pathway.
Lin28 (abnormal cell lineage)	Cold Shock RNA binding domain	Post transcriptional regulation	A cytoplasmic mRNA-binding protein that binds to enhance the translation of the Igf2 mRNA. Lin28 has also been shown to bind to the let-7 pre-miRNA and block production of the mature let-7 microRNA. Proposed to bind and regulates variety of miRNA and mRNA involved in differentiation.

DEDIFFERENTIATION

The more tempting scenario of ES cell applications in human health is the transplantation of cells, tissue and organs derived from these cells. However, cells or tissues differentiated from ES cells of one person will be immunologically incompatible to others. One way to overcome this problem could be patient specific or autologous pluripotent stem cells. However, there are no pluripotent stem cells that have been isolated from adults. One alternative to bypass this problem is to use easily available differentiated cells and dedifferentiated it into ES cells[26]. In this direction two approaches have been taken. In first approach, a nucleus from a differentiated cell was transferred to an undifferentiated enucleated unfertilized egg, which as a result dedifferentiated into totipotent stem cells. This process is known as 'nuclear reprogramming'. These manipulated eggs can be then transferred to surrogate mother to develop the animal. This technology was first used to make sheep clone popularly called "Dolly" and subsequently many other animals like mouse, pig, goat[27-30]. In second approach, dedifferentiation of differentiated cells was done by fusing adult cell with ES cells or growing adult cell in ES cell extract[31, 32]. These experiments were done, considering that unfertilized egg and ES cell have all the factors required to maintain them in totipotent and pluripotent conditions respectively. These dedifferentiation experiments explain that genome of differentiates cells can be remodeled and changed back in such a way that it can again attain same transcriptional profile as that of zygote or ES cells. The efficiency of these experiments is very low and they require sophisticated techniques. These limitations make it impractical to pursue these methods effectively for tissue engineering. Moreover, such manipulations may also lead to instability of the genome. These limitations always have restricted these techniques in laboratory stage.

What we learn however, from these experiments is that changes leading to dedifferentiation are epigenetic (epi 'in addition to' genetic) and does not involve any alteration in the DNA sequence. Epigenetic mechanisms are responsible for giving different genetic output in different cells using same genome. That is why our body has more than 220 types of cells whereas genome is same in all of them. Epigenetic mechanisms like DNA methylation, histone modifications (such as acetylation, methylation and ubiquitylation etc.), chromatin remodeling with chromatin-associated proteins, polycomb group proteins and transcription factors spatially and temporally regulate the transcriptome of a cell [33]. In principle, therefore, it should be possible to achieve greater efficiency in generating ES cells from any easily accessible differentiated cell of the body.

INDUCED PLURIPOTENT STEM (IPS) CELLS

As described in the previous section, reprogramming was demonstrated long back but essential components involved in the process remains unknown. Recent studies have identified number of factors and mechanisms responsible for differentiation and maintenance of 'stemness'. A major break through came when, differentiated adult mouse fibroblast cells were dedifferentiated in to ES cells by over expressing four crucial transcription factors, namely OCT4, SOX2 KLF4 and cMYC [34]. Such pluripotent stem cells were sorted by using Fbx15, a transcription factor and marker of pluripotency. This dedifferentiation of somatic cells into pluripotent stem cells is known as induced pluripotency. In this process terminally differentiated somatic cell is molecularly reprogrammed into pluripotent stem cells. Induced reprogramming also includes differentiation of cells from one lineage to another. For example, differentiation of B lymphocyte into macrophages by over expressing C/EBP alpha and C/EBP beta[35]. IPS cells show same morphology and growth properties as that of ES cells. Subcutaneous transplantation of these cells into nude mice resulted in tumors containing a variety of tissues from all three germ layers. However, the injection of iPS cells in blastocyst shows that they can participate in embryonic development, not in germ line transmission. These cells show very different epigenetic profile in comparison of ES cells[36]. Germ line transmission of iPS cells however was achieved by sorting pluripotent stem cells using Nanog gene expression instead of Fbx15[37]. These cells had almost similar transcriptional and epigenetic profile as compared to ES cells. More interestingly, when similar set of four transcription factors namely OCT4, SOX2 KLF4 and cMYC were over expressed in adult cells of human, they dedifferentiated into pluripotent cells[38]. In a different study induction of pluripotency in human adult somatic cells was achieved by over expressing Oct4, Sox2, Nanog and Lin28[39].

These studies suggest that the mechanism of pluripotency is highly inter-linked and that OCT4, SOX2 and NANOG are the essential factors. Epigenetic status, gene expression profile and telomerase level of iPS cells are very similar to naturally isolated ES cells, indicating a complete nuclear reprogramming of adult tissue into ES cells. Later on using the same set of transcription factors, many other cell types derived from brain, keratinocytes, blood, intestine, liver, stomach, pancreas, etc., were also reprogrammed into IPS cells demonstrating that, this method is universal and theoretically any cell type can be reprogrammed into iPS cells[40-42]. This technique was also successfully extended to produce iPS cells from body cells of ten different disease patient including Parkinson's disease and amylotropic lateral sclerosis [43]. With this series of achievements isolation of patient specific pluripotent stem cells has become possible. This has potential to be used in treatment of various human diseases (Fig 2). To make the scenario better, technical demands to isolate and use of iPS cells do not raise any ethical problems[44] and recently US food and drug administration (FDA) approved use of iPS cells to treat patient with spinal cord injury.

In the studies discussed here, over expression was done using virus based vectors[34, 37, 39]. The viral transfection systems used to insert the genes at random locations in the host genome as a cause of concern for potential therapeutic applications of these iPS cells. Such cells may carry mutations due to insertion in the genome, which may lead to problems, like tumor formation. During transgenesis using viral vectors, genomic viral elements may get active, which can lead to genome instability and cancer. In effort to avoid use of viral vectors, later it turned out that transient transfection of the key transcription factors is enough to reprogram the somatic cells[36, 45]. The reprogramming is possible in this case because transient activation of these transcription factors leads to activation of the endogenous genomic copy of these genes, which result into the reprogramming. Much more has to be done in this direction to achieve efficient and good quality of iPS cells in the cell culture conditions.

MECHANISM OF INDUCED PLURIPOTENCY

Induction of pluripotency by over expressing just four transcription factors challenged the complexity of the process involved in maintaining pluripotency. It looks surprisingly simple that by over expression of few factors leads to the changed transcriptome of the terminally differentiated cells leading to its dedifferentiation. Functions of these transcription factors were already studied in maintaining pluripotency as explained earlier in this chapter. Because induced pluripotency itself is very new technique, not much is known about the role of these factors in dedifferentiation. However, some factors and pathways involved in this processes are beginning to emerge [46]. As discussed above, two sets of four transcriptional factors (OCT4, SOX2, KLF4, cMYC and OCT4, SOX2, NANOG, LIN28) can induce pluripotency in differentiated cells[47]. Out of the first set, OCT4 and SOX2 are sufficient to induce true pluripotent stem cells with low efficiency, when Nanog was used as selection marker. They can bind to regulatory elements of the genes involved in pluripotency and differentiation at the same time to activate or suppress them, respectively. In differentiated cells these proteins possibly cannot access to the 'stemness genes' because of negative regulators and histone and DNA modifications. Upon over expression, these may activate chromatin modifiers and remodelers, which may inturn lead to activation of stemness genes. As KLF4 is a chromatin remodeler, it may facilitate and enhance this process, whereas cMYC is found to suppress the differentiated cell specific genes during early process of dedifferentiation by direct binding or activating repressors[48]. In the second set (OCT4, SOX2, NANOG, LIN28) LIN28 is dispensable but NANOG is a crucial and essential transcription factor required for pluripotency. NANOG recruits transcriptional activators to stemness genes for their activation[18, 49].

From these two sets of proteins, the most common are the OCT4, SOX2 and NANOG, which can be used to isolate iPS cells at higher efficiency. These three factors target thousands of loci in the differentiated cell to induced pluripotency. Over expression of this trio leads to epigenetic changes in the genome and at their own regulatory elements, which in tern changes the gene expression profile of the cell. Work has not been done extensively enough on the roles of all the six factors (OCT4, SOX2, NANOG, LIN28, KLF4 and cMYC) to dissect their independent functions in inducing pluripotency. It is possible that over expression of all these factors simultaneously may further enhance the efficiency of dedifferentiation but it needs to be tested.

The success in generating induced pluripotency from adult body cells is one of the biggest technological achievements of this decade[50]. It can be used to make patient specific stem cells without killing of embryos (Fig 2)[51]. This also eliminates immunological problems of allograph transplantation. In the near future individual specific iPS cells can be made to differentiate in desired lineage, which can be used as autograph. Such cells can also be used for drug screening and save patient from killing diseases.

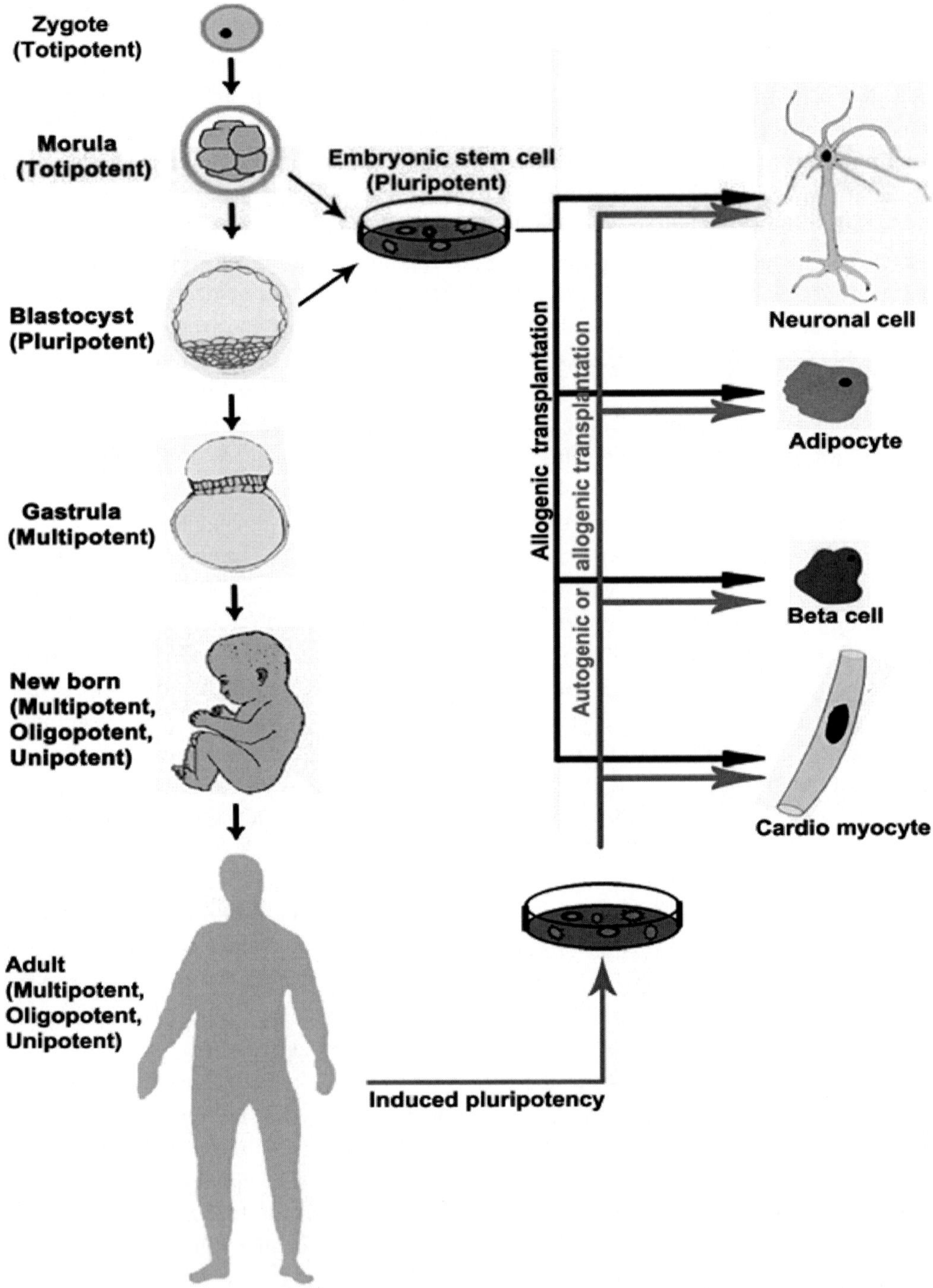

Figure 2. Isolation of pluripotent stem cells and their use: As zygote differentiates, it loses its potential to differentiate. Pluripotent stem cells are known to be present only at blastocyst stage during mammalian development. For the isolation of pluripotent stem cells from adult induced pluripotent technique can be used to dedifferentiate them into iPS cells.

Conclusion

Recent progress in understanding the establishment and maintenance of pluripotent ES cells has revealed the importance and functions of various key transcription factors and signaling pathways. Generation of iPS cells from mouse and human adult cells using same transcription factors suggest that the higher order mechanism of pluripotency is similar but their function and downstream signaling pathways may differ between the two species. Discovery of iPS cells opens up new possibility of using them in cell/tissue replacement therapy and as a model to study individual specific aspects of diseases. For using these cells, it is necessary to ensure that all the cells in the culture are true pluripotent and have all the genetic and epigenetic content intact for differentiation. It has been observed that sometimes pluripotent cells become resistant to differentiation. Use of impure iPS cells may lead to teretoma and tumor formation. In spite of enormous efforts already put in these directions, we still do not understand why and how pluripotent cells can differentiate in to all other cell types. To better understand these cells and their capabilities we need to understand the mechanism of induced pluripotency, relationship between the intracellular signaling pathways and the transcription factors involved this process. Understanding these aspects will lead to a truly regenerative medicine using the iPS cells.

Acknowledgment

We would like to thank Vijay Pratap Singh for valuable discussion, comments on the manuscript and providing ES cell phtograph. NPS acknowledges Council of Scientific and Industrial Research (CSIR) for research fellowship.

References

[1] Haskill J.S., Moore M.A., (1970). Two dimensional cell separation: comparison of embryonic and adult haemopoietic stem cells. *Nature, 226,* 853-854.

[2] Martin G.R., (1975). Teratocarcinomas as a model system for the study of embryogenesis and neoplasia. *Cell, 5,* 229-243.

[3] Martin G.R., (1980). Teratocarcinomas and mammalian embryogenesis. *Science, 209,* 768-776.

[4] Bradley A., Evans M., Kaufman M.H., Robertson E., (1984). Formation of germ-line chimaeras from embryo-derived teratocarcinoma cell lines. *Nature, 309,* 255-256.

[5] Thomson J.A., Itskovitz-Eldor J., Shapiro S.S. et al., (1998). Embryonic stem cell lines derived from human blastocysts. *Science, 282,* 1145-1147.

[6] Li P., Tong C., Mehrian-Shai R. et al., (2008). Germline competent embryonic stem cells derived from rat blastocysts. *Cell, 135,* 1299-1310.

[7] Thomson J.A., Kalishman J., Golos T.G. et al., (1995). Isolation of a primate embryonic stem cell line. *Proc Natl Acad Sci U S A, 92,* 7844-7848.

[8] Lee H.W., Blasco M.A., Gottlieb G.J., Horner J.W., 2nd, Greider C.W., DePinho R.A., (1998). Essential role of mouse telomerase in highly proliferative organs. *Nature, 392,* 569-574.

[9] Escary J.L., Perreau J., Dumenil D., Ezine S., Brulet P., (1993). Leukaemia inhibitory factor is necessary for maintenance of haematopoietic stem cells and thymocyte stimulation. *Nature, 363,* 361-364.

[10] Okita K., Yamanaka S., (2006). Intracellular signaling pathways regulating pluripotency of embryonic stem cells. *Curr Stem Cell Res Ther, 1,* 103-111.

[11] Feng G.S., (2007). Shp2-mediated molecular signaling in control of embryonic stem cell self-renewal and differentiation. *Cell Res, 17,* 37-41.

[12] Ying Q.L., Nichols J., Chambers I., Smith A., (2003). BMP induction of Id proteins suppresses differentiation and sustains embryonic stem cell self-renewal in collaboration with STAT3. *Cell, 115,* 281-292.

[13] Matzuk M.M., Kumar T.R., Bradley A., (1995). Different phenotypes for mice deficient in either activins or activin receptor type II. *Nature, 374,* 356-360.

[14] Amit M., Carpenter M.K., Inokuma M.S. et al., (2000). Clonally derived human embryonic stem cell lines maintain pluripotency and proliferative potential for prolonged periods of culture. *Dev Biol, 227,* 271-278.

[15] Mishra L., Derynck R., Mishra B., (2005). Transforming growth factor-beta signaling in stem cells and cancer. *Science, 310,* 68-71.

[16] Kitisin K., Saha T., Blake T. et al., (2007). Tgf-Beta signaling in development. *Sci STKE, 2007,* cm1.

[17] Watabe T., Miyazono K., (2009). Roles of TGF-beta family signaling in stem cell renewal and differentiation. *Cell Res, 19,* 103-115.

[18] Chambers I., Silva J., Colby D. et al., (2007). Nanog safeguards pluripotency and mediates germline development. *Nature, 450,* 1230-1234.

[19] Xiao L., Yuan X., Sharkis S.J., (2006). Activin A maintains self-renewal and regulates fibroblast growth factor, Wnt, and bone morphogenic protein pathways in human embryonic stem cells. *Stem Cells, 24,* 1476-1486.

[20] Avilion A.A., Nicolis S.K., Pevny L.H., Perez L., Vivian N., Lovell-Badge R., (2003). Multipotent cell lineages in early mouse development depend on SOX2 function. *Genes Dev, 17,* 126-140.

[21] Mitsui K., Tokuzawa Y., Itoh H. et al., (2003). The homeoprotein Nanog is required for maintenance of pluripotency in mouse epiblast and ES cells. *Cell, 113,* 631-642.

[22] Pan G.J., Chang Z.Y., Scholer H.R., Pei D., (2002). Stem cell pluripotency and transcription factor Oct4. *Cell Res, 12,* 321-329.

[23] Yeom Y.I., Fuhrmann G., Ovitt C.E. et al., (1996). Germline regulatory element of Oct-4 specific for the totipotent cycle of embryonal cells. *Development, 122,* 881-894.

[24] Tay Y., Zhang J., Thomson A.M., Lim B., Rigoutsos I., (2008). MicroRNAs to Nanog, Oct4 and Sox2 coding regions modulate embryonic stem cell differentiation. *Nature, 455,* 1124-1128.

[25] D'Amour K.A., Agulnick A.D., Eliazer S., Kelly O.G., Kroon E., Baetge E.E., (2005). Efficient differentiation of human embryonic stem cells to definitive endoderm. *Nat Biotechnol, 23,* 1534-1541.

[26] Surani M.A., Barton S.C., Norris M.L., (1986). Nuclear transplantation in the mouse: heritable differences between parental genomes after activation of the embryonic genome. *Cell, 45,* 127-136.

[27] Wilmut I., Schnieke A.E., McWhir J., Kind A.J., Campbell K.H., (1997). Viable offspring derived from fetal and adult mammalian cells. *Nature, 385,* 810-813.

[28] Wakayama T., Perry A.C., Zuccotti M., Johnson K.R., Yanagimachi R., (1998). Full-term development of mice from enucleated oocytes injected with cumulus cell nuclei. *Nature, 394,* 369-374.

[29] Baguisi A., Behboodi E., Melican D.T. et al., (1999). Production of goats by somatic cell nuclear transfer. *Nat Biotechnol, 17,* 456-461.

[30] Polejaeva I.A., Chen S.H., Vaught T.D. et al., (2000). Cloned pigs produced by nuclear transfer from adult somatic cells. *Nature, 407,* 86-90.

[31] Cowan C.A., Atienza J., Melton D.A., Eggan K., (2005). Nuclear reprogramming of somatic cells after fusion with human embryonic stem cells. *Science, 309,* 1369-1373.

[32] Surani M.A., (2005). Nuclear reprogramming by human embryonic stem cells. *Cell, 122,* 653-654.

[33] Boyer L.A., Plath K., Zeitlinger J. et al., (2006). Polycomb complexes repress developmental regulators in murine embryonic stem cells. *Nature, 441,* 349-353.

[34] Takahashi K., Yamanaka S., (2006). Induction of pluripotent stem cells from mouse embryonic and adult fibroblast cultures by defined factors. *Cell, 126,* 663-676.

[35] Xie H., Ye M., Feng R., Graf T., (2004). Stepwise reprogramming of B cells into macrophages. *Cell, 117,* 663-676.

[36] Okita K., Nakagawa M., Hyenjong H., Ichisaka T., Yamanaka S., (2008). Generation of mouse induced pluripotent stem cells without viral vectors. *Science, 322,* 949-953.

[37] Okita K., Ichisaka T., Yamanaka S., (2007). Generation of germline-competent induced pluripotent stem cells. *Nature, 448,* 313-317.

[38] Takahashi K., Tanabe K., Ohnuki M. et al., (2007). Induction of pluripotent stem cells from adult human fibroblasts by defined factors. *Cell, 131,* 861-872.

[39] Yu J., Vodyanik M.A., Smuga-Otto K. et al., (2007). Induced pluripotent stem cell lines derived from human somatic cells. *Science, 318,* 1917-1920.

[40] Zaehres H., Scholer H.R., (2007). Induction of pluripotency: from mouse to human. *Cell, 131,* 834-835.

[41] Park I.H., Zhao R., West J.A. et al., (2008). Reprogramming of human somatic cells to pluripotency with defined factors. *Nature, 451,* 141-146.

[42] Zhao R., Daley G.Q., (2008). From fibroblasts to iPS cells: induced pluripotency by defined factors. *J Cell Biochem, 105,* 949-955.

[43] Soldner F., Hockemeyer D., Beard C. et al., (2009). Parkinson's disease patient-derived induced pluripotent stem cells free of viral reprogramming factors. *Cell, 136,* 964-977.

[44] Loring J.F., Campbell C., (2006). Science and law. Intellectual property and human embryonic stem cell research. *Science, 311,* 1716-1717.

[45] Woltjen K., Michael I.P., Mohseni P. et al., (2009). piggyBac transposition reprograms fibroblasts to induced pluripotent stem cells. *Nature.*

[46] Niwa H., (2007). How is pluripotency determined and maintained? *Development, 134,* 635-646.

[47] Jaenisch R., Young R., (2008). Stem cells, the molecular circuitry of pluripotency and nuclear reprogramming. *Cell, 132,* 567-582.

[48] Sridharan R., Tchieu J., Mason M.J. et al., (2009). Role of the murine reprogramming factors in the induction of pluripotency. *Cell, 136,* 364-377.

[49] Pan G., Thomson J.A., (2007). Nanog and transcriptional networks in embryonic stem cell pluripotency. *Cell Res, 17,* 42-49.

[50] Maherali N., Hochedlinger K., (2008). Guidelines and techniques for the generation of induced pluripotent stem cells. *Cell Stem Cell, 3,* 595-605.

[51] Yamanaka S., (2007). Strategies and new developments in the generation of patient-specific pluripotent stem cells. *Cell Stem Cell, 1,* 39-49.

In: Recent Trends in Biotechnology, Volume 1
Editors: M.P. Singh, A. Agrawal and B. Sharma

ISBN 978-1-60876-148-7

Chapter 8

STEM CELL RESEARCH: PROSPECTS AND RETROSPECTS

Devendra Kumar Rai[1], Anju Agarwal[*] and B. Sharma[1†]

[1] Department of Biochemistry, University of Allahabad
Allahabad-211002, India

[2] Department of Zoology, SNSenBV PG College,
CSJM Kanpur University, India

ABSTRACT

Stem cells are cells found in most of the multi-cellular organisms. They are characterized by the ability to renew themselves through mitotic cell division and differentiating into a diverse range of specialized cell types. The prospect of this technology is immense despite the fact that research in the stem cell segment has been exhaustively discussed and criticized at numerous political, ethical and public forums in several countries. It has raised a hope among the scientist community about the possible use of stem cells as a tool not only to replace parts for failing organs but also to treat certain debilitating diseases like diabetes and Parkinson's disease. Most cells within an animal or human body are committed in fulfilling a specific function within the body. In contrast, stem cells are unique and important set of cells that are not specialized. These cells retain the ability to differentiate into many or all of the different cell types in the body and thereby play a critical role in repairing organs and body tissues throughout life. They can therapeutically divide without limit to replenish other cells as long as the person or animal is alive. When a stem cell divides, each new cell has the potential to either remain a stem cell or become another type of cell with a more specialized function, such as a muscle cell, red blood cell, or brain cell. Apart from application of stem cells in regenerative therapy, it also helps in understanding the early human development and the pathophysiology of genetic diseases at the cellular level. However, there are some technical problems and ethical concerns that need to be addressed before applying stem cells into clinical practice. The present chapter describes some salient features, current scenario and future prospects / retrospect of clinical applications of stem cells.

---

[*] E-mail:anjuaa@rediffmail.com

[†] Corresponding author: E-mail: sharmabi@yahoo.com

INTRODUCTION

Stem cells are characterized by their ability to renew themselves through mitotic cell division and differentiating into a diverse range of specialized cell types. Research in the stem cell field grew out of findings by Canadian scientists Ernest A. McCulloch and James E. Till in the 1960s [1,2]. Recent advancements in biomedical research have revolutionized a spurt of medical facilities, which not only help in better understanding of various diseases and disorders but also provide sophisticated diagnostic measures and therapies. Stem cell research is the recent and one of the most promising ones in this area. It promises to provide miraculous cure to degenerative diseases, which could otherwise be delayed in earlier methods of treatment. It can also be very helpful in understanding various developmental cases which can further be exploited in deciding the developmental fate of target cells, tissues and organs. The main clinical application of stem cells is to use them as a source of donor cells to replace diseased cells in transplantation therapy.

A chronological perspective of stem cell research has been displayed in Table 1. Ever since its advent, the stem cell research has been a center of controversies despite its unlimited promises in improving the health status of mankind. Especially after the outburst of the fictitious work of the South Korean scientist, Hwang Woo Suk on cloned human embryos and stem cells, stem cell research has forced the global scientific fraternity to be more careful while dealing with such an important issue. Very recently, The Nobel Assembly of Karolinska Institute of Sweden jointly awarded the Nobel Prize in Physiology or Medicine for 2007 to Drs. Mario R. Capecchi (USA), Martin J. Evans (UK) and Oliver Smithies (USA) for their discoveries of "principles for introducing specific gene modifications in mice by the use of embryonic stem cells". In brief, their discoveries led to the creation of an immensely powerful technology referred to as ´gene targeting in mice`, which is discussed later. It is now being applied to virtually all areas of biomedicine – from basic research to the development of new therapies. Gene targeting is often used to inactivate single genes. Such gene "knockout" experiments have elucidated the roles of numerous genes in embryonic development, adult physiology, aging and disease. To date, more than ten thousand mouse genes (approximately half of the genes in the mammalian genome) have been knocked out. Ongoing international efforts will make "knockout mice" for all genes available within the near future. With gene targeting it is now possible to produce almost any type of DNA modification in the mouse genome, allowing scientists to establish the roles of individual genes in health and disease. Gene targeting has already produced more than five hundred different mouse models of human disorders, including cardiovascular and neuro-degenerative diseases, diabetes and cancer.

Stem cells are undifferentiated parental cells of all cells of the body. They are capable of multiplying and replenishing themselves for long periods. However, upon receiving appropriate signals they can divide and grow into any of the specialized cell types of the body and therefore can make up organs like limbs, lung, liver, brain, pancreas etc. This property of stem cells termed as "pluripotency" holds key to their potential application in regenerative medicine and in studying developmental processes. The different types of stem cells based on their hierarchical and differentiation potential are enlisted in Box 1. One of the representative stem cells is shown in Figure 1 [3].

Table 1. Chronology of key events in stem cell research

Dates	Events/Discoveries	References
1908	The term "stem cell" was proposed by Russian histologist, Alexander Maksimov (1874–1928) at congress of hematologic society in Berlin. It postulated existence of haematopoietic stem cells.	
1960	Joseph Altman and Gopal Das presented scientific evidence of adult neurogenesis, ongoing stem cell activity in the brain; their reports contradict Cajal's "no new neurons" dogma and are largely ignored	
1963	McCulloch and Till illustrate the presence of self-renewing cells in mouse bone marrow	
1968	Bone marrow transplant between two siblings successfully treats SCID.	
1978	Haematopoietic stem cells are discovered in human cord blood	
1981	Mouse embryonic stem cells are derived from the inner cell mass by scientists Martin Evans, Matthew Kaufman, and Gail R. Martin. Gail Martin is attributed for coining the term "Embryonic Stem Cell".	
1992	Neural stem cells are cultured *in vitro* as neurospheres	
1997	Leukemia is shown to originate from a haematopoietic stem cell, the first direct evidence for cancer stem cells	
1998	James Thomson and coworkers derive the first human embryonic stem cell line.	[14]
2000-	Several reports of adult stem cell plasticity are published.	
2001	Scientists at Advanced Cell Technology clone first early (four- to six-cell stage) human embryos for the purpose of generating embryonic stem cells	[15]
2003	Dr. Songtao Shi of NIH discovers new source of adult stem cells in children's primary teeth.	[16]
2004-05	Korean researcher Hwang Woo-Suk claims to have created several human embryonic stem cell lines from unfertilised human oocytes. The lines were later shown to be fabricated.	
2005	Researchers claimed discovery of third category of stem cell, dubbed cord-blood-derived embryonic-like stem cells (CBEs), derived from umbilical cord blood, which were able to differentiate into more types of tissue than adult stem cells	
Aug.2006	Rat Induced pluripotent stem cells were investigated.	[17]
Oct.2006	Scientists created the first ever artificial liver cells using umbilical cord blood stem cells.	
Jan.2007	Scientists reported discovery of a new type of stem cell in amniotic fluid, providing an alternative to embryonic stem cells for use in research and therapy.	[18]

Table 1. Continued

Dates	Events/Discoveries	References
June 2007	Research reported that normal skin cells can be reprogrammed to an embryonic state in mice. Shoukhrat Mitalipov reported first successful creation of a primate stem cell line through somatic cell nuclear transfer.	[19,20]
Oct.2007	Mario Capecchi, Martin Evans, and Oliver Smithies win the 2007 Nobel Prize for Physiology or Medicine for their work on embryonic stem cells from mice using gene targeting.	
Nov.2007	Human induced pluripotent stem cells	[21,22]
Jan.2008	First human embryonic stem cells without destruction of the embryo was created. Development of human cloned blastocysts following somatic cell nuclear transfer with adult fibroblasts was made.	[23,24]
Feb 2008	Generation of Pluripotent Stem Cells from adult mouse liver and Stomach.	
March 2008	The first published study of successful cartilage regeneration in the human knee using autologous adult mesenchymal stem cells came out.	[25]
Oct.2008	Pluripotent stem cells from spermatogonial cells of adult human testis by culturing the cells *in vitro* under leukemia inhibitory factor (LIF) supplementation and Embryonic-like stem cells from a single human hair were produced.	[26,27]
March 2009	Discovery of a way to produce embryonic-like stem cells from normal adult cells by using a novel "wrapping" procedure to deliver specific genes to adult cells to reprogram them into stem cells without the risks of using a virus to make the change.The use of electroporation is said to allow for the temporary insertion of genes into the cell. Discovery of a way to improve chemotherapy of mouse muscle stem cells was done.US President Obama lifted federal funding limits on human embryonic instituted by former President Bush	[28,29]

Embryonic stem cells are considered pluripotent instead of totipotent because they do not have the ability to become part of the extra-embryonic membranes of the placenta.

Unlike other cells of the body all stem cells, regardless of their source, have some unique properties, which make them a special type of cell with tremendous therapeutic applications. They are capable of dividing and renewing themselves for long periods. They are unspecialized so they can choose to become one of the many different types of cells present in the body based on signals they receive from their environments.

Box 1. Stem cells types (on the basis of hierarchy and differentiation potential)

Not all stem cells are created equal. The potential for stem cells to differentiate into other specialized cells changes as an embryo develops. Stem cells are categorized by potency which describes the specificity of the cell.

1. Totipotent stem cells are produced from the fusion of an egg and sperm cell. Cells produced by the first few divisions of the fertilized egg cell are also totipotent. These cells can grow into any type of cell without exception. This means that these cells can form even those cells necessary for an embryo to develop into a human, such as placenta cells. Examples are the zygote formed at egg fertilization and the first few cells that result from the division of the zygote.
2. Pluripotent stem cells are the descendants of totipotent cells. Several days after conception, as an embryo begins to develop, its stem cells become pluripotent. Such cells can form any kind of cell found in adults, making them very versatile. Unlike totipotent stem cells, however, they can't become cells necessary for an embryo to develop into a human. Cells with similar properties are also found in fetal tissue. Examples include embryonic stem cells and cells that are derived from the mesoderm, endoderm, and ectoderm germ layers that are formed in the beginning stages of embryonic stem cell differentiation.
3. Multipotent stem cells can produce only cells of a closely related family of cells (e.g. blood cells such as red blood cells, white blood cells and platelets). Eventually, pluripotent stem cells become multipotent stem cells; the type found in adults. Multipotent stem cells can no longer develop into all or most cell types. Instead, they can develop into certain cell types within a specific tissue, organ or system. For example, bone marrow stem cells can develop into all types of blood cells, which are normally produced by our bone marrow, as well as bone, cartilage and fat cells. Stem cells in our brain can turn into nerve cells and other types of brain cells. Examples include hematopoietic (adult) stem cells that can become red and white blood cells or platelets.
4. Oligopotent stem cells have the ability to differentiate into a few cells. Examples include (adult) lymphoid or myeloid stem cells.
5. Unipotent stem cells have the ability to only produce cells of their own type, but have the property of self-renewal required to be labed a stem cell. Examples include (adult) muscle stem cells.

Most of the specialized cells in our body (such as cells of muscle, blood, or nerve) do not replicate themselves. The supply of these cells is maintained by stem cells, which after repeated replication differentiate into required specialized cells. In this way, cells are continuously replenished as they die thereby maintaining homeostasis. When cells replicate themselves many times over it is called proliferation. A starting population of stem cells that proliferates for many months in the laboratory can yield hundreds of millions of cells. If the resulting cells are identical (in terms of being unspecialized themselves and capable of generating specialized cells) to the parent stem cells, the stem cells are said to be capable of long-term self-renewal.

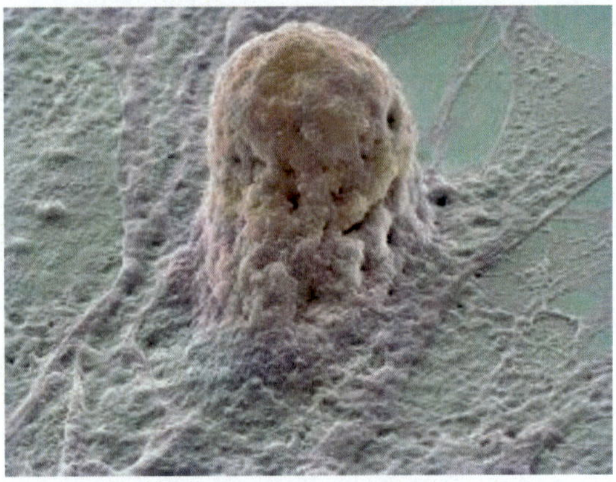

Reproduced from: http://www.rte.ie/news /2007/1008/nobel.html

Figure 1. A representative stem cell.

Differentiation is the process; whereby stem cells give rise to specialized cells i.e. undifferentiated stem cells differentiate into the different cell types upon receiving appropriate signals from their adjoining environment. These signals include chemicals secreted by other cells, physical contact with neighbouring cells, and contact with molecules in the microenvironment.

Until recently, adult stem cells were considered to be irreversibly committed to specific lineages of differentiation. Hematopoietic Stem Cells (HSCs), for example, normally give rise to all types of blood cells such as red blood cells, white blood cells and platelets. It was previously believed that HSCs could not give rise to any cells of a different tissue. However, a number of experiments over the last several years have raised the possibility that stem cells from one tissue may be able to give rise to cell types of a completely different tissue. This phenomenon is known as 'stem cell plasticity'. Examples of such plasticity include bone marrow stem cells becoming neurons, or pancreatic islet cells that are capable of producing insulin. Exploring the possibility of using adult stem cells for cellular therapies has recently become an active area of research. The apparent plasticity of adult stem cells has forced scientists to reconsider many fundamental concepts of stem cell biology.

On the basis of their potential to differentiate into other types of cells, the stem cells have been classified into two major types i.e. the stem cells based on their origin (1) Adult stem cells (ASCs) and (2) Embryonic stem cells (ESCs). The adult stem cells (ASCs) are the undifferentiated (unspecialized) stem cells residing in differentiated or specialized tissues or organs and replace the cells lost in normal wear and tear, injury and diseases. The ASCs usually divide to generate progenitor or precursor cells, which then differentiate or develop into mature cell types that have characteristic shapes and specialized functions, for example muscle cell contraction or nerve cells signaling. Sources of adult stem cells include bone marrow, blood, the cornea and retina of eye, brain, skeletal muscle, dental pulp, liver, skin, the lining of gastrointestinal tract and pancreas. However, under the influence of certain biochemical signals they can develop into the cells of an altogether different tissue, for example stem cells of bone marrow can develop into blood cells and muscle cells etc. (Figure 2). However this capacity is limited and is known as multipotency.

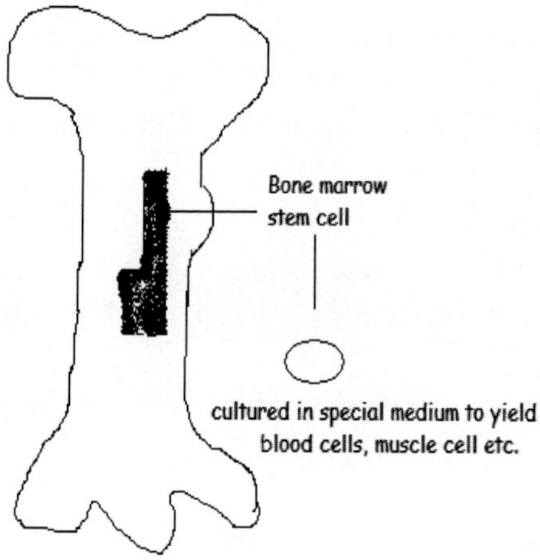

Figure 2. Adult stem cells and their differentiation (such as bone marrow stem cells).

The embryonic stem cells (ESCs), on the other hand, are extracted from inner cell mass of blastocyst stage (5-7days) embryo. These embryos are obtained from *in vitro* fertility centers. The ESCs can develop into any cell type in the body. This property is called pluripotency. Pluripotency is the hallmark of ESCs and basis of their much-promised future application in biomedical therapeutics. Embryonic stem cells must be obtained when an embryo is in its early developmental stage, that is, when the fertilized egg has divided to form about 1000 cells. These cells are separated and maintained in a cell culture dish, thereby halting embryonic development towards creating an individual. Depending on the need to generate a particular type of tissue, the cells are inoculated into specially designed broths (Figure 3).

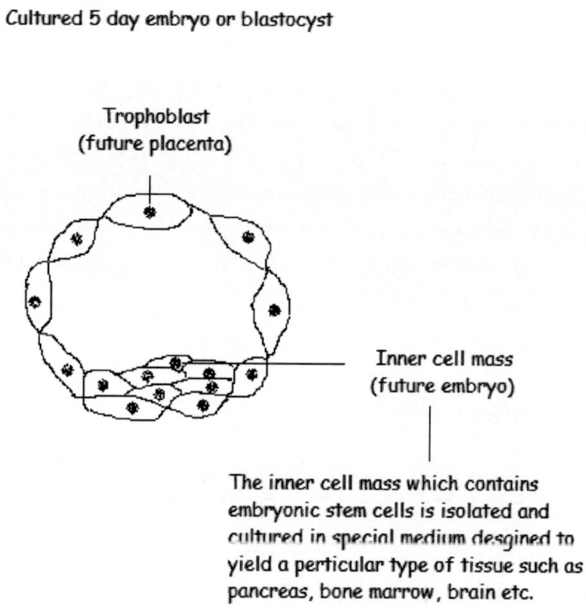

Figure 3. Embryonic stem cells (ESCs).

Apart from the above two populations, stem cells can also be derived from primordial germ cells in the gonadal ridge of 5-10 week old fetus, known as primordial or embryonic germ cells (EGCs, Figure 4). The discarded umbilical cord, aborted fetus and placenta are among other rich sources of stem cells. However, the stem cells derived from these sources are less versatile and pluripotent than ESCs.

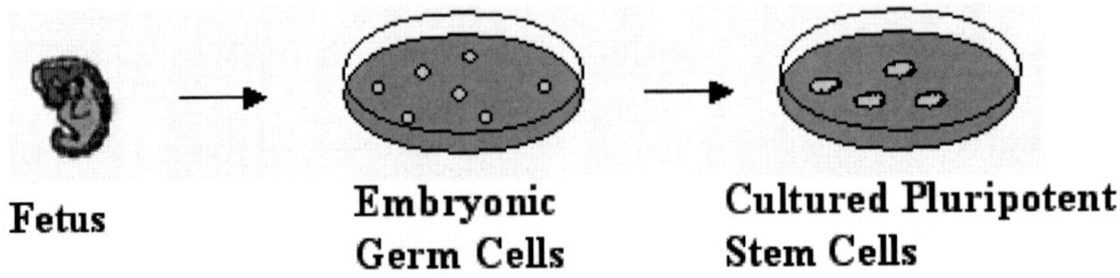

Figure 4. Derivation of stem cells from embryonic germ cells cells (EGCs).

There are many other types of stem cells as well:

(I) Hematopoietic Stem Cells

The hematopoietic stem cells (HSCs) are those from which all red and white blood cells evolve. They are found in the bone marrow, where all the blood cells are produced (about 2 to 3 million per second). A number of experiments have suggested that certain adult stem cell types are pluripotent. They show that hematopoietic stem cells may differentiate into three major types of brain cells (neurons, oligodendrocytes, and astrocytes); skeletal muscle cells; cardiac muscle cells; and liver cells. Haematopoiesis is the process whereby mature blood cells of distinct lineages are produced from mulitpotent haematopoietic stem cells (HSCs). Recently Cumano et al. [4] have shown that in mice, HSCs are generated in an intraembryonic region, termed the splanchnopleura (Sp), and not from the extraembryonic yolk sac (YS) blood islands. During vertebrate embryogenesis, haematopoietic cells first appear within the extraembryonic YS. As the bone marrow and fetal liver (the main haematopoietic organs in mammals) require an input of exogenous haematopoietic precursors to generate differentiated progeny, it was suggested that HSCs originate in the YS and later migrate to the fetal liver. However, the intraembryonic Sp region has also been shown to have haematopoietic activity. The only cells capable of adult long-term haematopoiesis are from the Sp region. So, during mouse embryogenesis, HSCs are generated intraembryonically and do not derive from the YS [4].

(II) Mesenchymal Stem Cells

The mesenchymal stem cells (MSCs), also known as bone marrow stromal cells, are found in bone marrow. They are multipotent stem cells that can differentiate into a variety of cell types. Cell types that MSCs have been shown to differentiate into *in vitro* or *in vivo* into a variety of cell types which include osteoblasts (bone cells), chondrocytes (cartilage cells), myocytes, adipocytes (fat cells), connective tissue cells such as those in tendons and, as described lately, beta-pancreatic islets cells [5]. Experiments have shown that mesenchymal stem cells may also differentiate into cardiac muscle cells and skeletal muscle cells.

While the terms Mesenchymal Stem Cell (MSCs) and Marrow Stromal Cell (MSCs) have been used interchangeably, neither term is sufficiently descriptive. Mesenchyme is embryonic connective tissue that is derived from the mesoderm and that differentiates into hematopoietic and connective tissue, whereas MSCs do not differentiate into hematopoietic cells. Some of the specific properties of MSCs include:

(i) morphologically they look like a small cell body with a few cell processes that are long and thin [6,7]
(ii) having tremendous capacity for differentiation/self-renewal while maintaining their multipotency [8]
(iii) avoid allorecognition, interfere with dendritic cell and T-cell function and generate a local immunosuppressive microenvironment by secreting cytokines [9] and
(iv) the mesenchymal stem cells can be activated and mobilized if needed. However, the efficiency is very low. For instance, damage to muscles heals very slowly. However, if there were a method of activating the mesenchymal stem cells then such wounds would heal much faster.

(III) Endothelial Stem Cells

The endothelial stem cells (EndSCs) are multipotent stem cells. They are one of the three types of stem cells to be found in bone marrow and are derived from the endoderm, which is the inside layer of the cells from the inner cell mass of the blastocyst [10]. The endodermal stem cells found there gives rise to lungs, other respiratory structures, digestive tract, liver and pancreas, or generally "the gut".

(IV) Neural Stem Cells

Neural stem cells (NSCs) are found in adult neural tissue, the brain. Neural stem cells exist not only in the developing mammalian nervous system but also in the adult nervous system of all mammalian organisms, including humans. Neural stem cells can also be derived from more primitive embryonic stem cells. The location of the adult stem cells and the brain regions to which their progeny migrate in order to differentiate remain unresolved, although the number of viable locations is limited in the adult. The mechanisms that regulate endogenous stem cells are poorly understood. Potential uses of stem cells in repair include transplantation to repair missing cells and the activation of endogenous cells to provide "self-repair." [11]. They generate the brain's three major cell types-neurons, or nerve cells, and astrocytes and oligodendrocytes, which are non-neuronal cells. These non-neuronal cells are called glial (supporting) cells. Experiments have shown that brain stem cells may also differentiate into blood cells and skeletal muscle cells.

Stem cells can be obtained from several sources such as:

(i) Spare embryos: stem cells can come from left over embryos stored at fertility clinics that were not used by couples to have children
(ii) Special purpose embryos: embryos are created by *in vitro* fertilization (artificially in the lab) for the sole purpose of extracting their stem cells
(iii) Cloned embryos: embryos are cloned in labs using somatic nuclear transfer method in order to harvest their stem cells
(iv) Aborted fetuses: stem cells are taken from fetuses in early development that have been aborted
(v) Umbilical cords: after-childbirth tissues from the umbilical cords hold potential for research
(vi) Adult tissues or organs: stem cells are obtained from the tissues or organs of living adults during surgery
(vii) Cadavers: isolation and survival of neural progenitor cells from human post-natal or post-mortem tissues (up to 20 h after death) have been reported which provides an additional source of human stem cells.

Stem cells can be used in a variety of ways in basic research, clinical research and therapeutics. By studying development of stem cells into different cell types, the genes involved or the specific pathways leading to differentiation of a particular cell type can be elucidated. ES cell studies can also give an idea about some of the serious medical disorders like cancer; birth defects etc. whether these are due to chromosomal cell division or differentiation. Better knowledge of these diseases may lead to the molecular clues and help in designing new strategies of therapy. Further, the ESCs can also be used in testing drugs.

However, the most important application of stem cells is the generation of new cell types and tissues, which are used in cell-based regenerative therapies. Stem cells have already been in practice for treating various disorders in which stem cells from the patient's body itself is taken, cultured and induced by appropriate signals. When sufficient number of these cells is obtained, it is used to replace the damaged tissues. These stem cells colonize in the particular tissue of interest and repair the damaged region. This approach has got some serious limitations such as ASCs used in this type of therapy possess very poor proliferating capacity in the tissue culture conditions and can develop into very few cell types only. By virtue of being isolated from patient's own body ASCs avoid risk of immune rejection posed by host body (due to genetic factors) induced by appropriate signals. Among the most commonly cited potential applications are, treatment of fractures, burns and other injuries, as well as developing effective therapies for a range of serious degenerative conditions such as Parkinson's, Alzheimer's, diabetes, and certain types of heart disease.

Limbal cord stem cells have been used in the cornea transplants. Such diseases are major causes of illness and death. For many of them, no effective treatment currently exists. At present the most widely used stem cell-based therapy is bone marrow transplant. Future regenerative therapies could transplant tissues containing stem cells to replace diseased or damaged tissues. It is also likely that therapies will be developed that use stem cells purified and cultured in the laboratory. These could be administered to patients as undifferentiated cells. Alternatively, the stem cells could be differentiated to cells of the appropriate type in the laboratory and then used for therapy. A final possibility is application of transplanting tissues made in the laboratory using differentiated stem cells. While there is widespread agreement over the potential therapeutic benefits of stem cells, further research is required before such therapies become a reality. A concise list of possible application of stem cells in therapeutics is presented in Box 2.

Box 2. Possible uses of tissue derived from stem cells to treat diseases

Cell Type	Target Disease
Blood cells	Cancer, Immunodeficiencies, Inherited blood diseases, Leukemia
Bone cells	Osteoporosis
Cartilage cells	Osteoorthritis
Heart Muscle cells	Heart attacks, Congestive heart failures
Insulin producing cells	Diabetes
Liver cells	Hepatitis, Cirrhosis
Neural (Nerve) Cells	Stroke, Parkinson's disease, Alzheimer's disease, Spinal cord injury, Multiple sclerosis
Retinal (eye) cells	Macular degeneration
Skeletal muscle	Muscular dystrophy
Skin cells	Burns, wound healing

The regeneration of tissues and organs from stem cells is probably the most important possible application of stem cell research. Currently, organs are donated and transplanted, but the demand for organs far exceeds the supply. Stem cells could potentially be used to grow a particular type of tissue or organ if directed to differentiate in a certain way. Stem cells that lie just beneath the skin, for example, have been used to engineer new skin tissue that can be grafted on to burn victims. Additionally, replacement of cells and tissues may be used to treat diseases associated to brain such as Parkinson's and Alzheimer's by replenishing damaged tissue, bringing back the specialized brain cells. Embryonic stem cells have recently been directed to differentiate into these types of cells making treatments so promising.

Healthy heart cells developed in a laboratory may one day be transplanted into patients with heart disease, repopulating the heart with healthy tissues. Similarly, people with type I diabetes may receive pancreatic cells to replace the ineffective insulin-producing cells that have been lost or destroyed by the patient's own immune system with the fresh functional cells. The only current therapy is a pancreatic transplant, and it is unlikely to occur due to a small supply of pancreases available for transplant.

Adult hematopoietic stem cells found in blood and bone marrow have been used for years to treat blood related diseases such as leukemia, sickle cell anemia, and other immunodeficiency complications. These cells are capable of producing all blood cell types, such as red blood cells that carry oxygen to white blood cells that fight disease. Difficulties arise in the extraction of these cells through the use of invasive bone marrow transplants. However, hematopoietic stem cells have also been found in the umbilical cord and placenta. This has led some scientists to call for an umbilical cord blood bank to make these powerful cells more easily obtainable and to decrease the chances of a body's rejecting therapy.

Stem cell research is also useful for learning about human development. Undifferentiated stem cells eventually differentiate partly because a particular gene is turned on or off. Stem cell researchers may help to clarify the role that genes play in determining what genetic traits or mutations they receive. Cancer and other birth defects are also affected by abnormal cell division and differentiation. New therapies for diseases may be developed if we better understand how these agents attack the human body. Another reason why stem cell research is being pursued is to develop new drugs. Scientists could measure a drug's effect on healthy, normal tissue by testing the drug on tissue grown from stem cells rather than testing the drug on human volunteers.

In India owing to poor infrastructure, low life expectancy and most ironically the poor distribution of medical facilities, patients come to the clinics during the most developed stage of the disease, when it has progressed to an irreversible extent. Thus the role of stem cell based regenerative therapy becomes extremely important.

In this context, it is important to mention about the work of Nobel Laureates for Physiology or Medicine (2007) [12], which is associated with the modification of genes by homologous recombination (a process of exchange of DNA sequences within chromosome pairs increasing variability) and stem cells applications. Mario Capecchi and Oliver Smithies both were of the view that homologous recombination could be used to specifically modify genes in mammalian cells. Capecchi demonstrated that homologous recombination could take place between introduced DNA and the chromosomes in mammalian cells. He showed that defective genes could be repaired by homologous recombination with the incoming DNA. Smithies initially tried to repair mutated genes in human cells with the hope that certain inherited blood diseases could be treated by correcting the disease-causing mutations in bone marrow stem cells. He also discovered that endogenous genes could be targeted irrespective of their activity and suggested that all genes may be accessible to modification by homologous recombination.

Capecchi and Smithies isolated another cell type to create gene-targeted animals. They needed a cell type, which could give rise to germ cells to facilitate the inheritance of the DNA modifications. Evans had worked with mouse embryonal carcinoma (ECs) cells, which could give rise to almost any cell type. He had the vision to use EC cells as vehicles to introduce genetic material into the mouse germ line. His attempts were initially unsuccessful because EC cells carried abnormal chromosomes and could not therefore contribute to germ cell formation. Evans discovered as an alterante that chromosomally normal cell cultures could be established directly from early mouse embryos. These cells are now referred to as embryonic stem cells (ESCs) .

In the next step, they showed that ESCs could contribute to the germ line. It was done by injecting embryos from one mouse strain with ESCs from another mouse strain. These mosaic embryos (i.e. composed of cells from both strains) were then carried by surrogate mothers. The mosaic offspring was subsequently mated, and the presence of ESCs-derived genes detected in the pups. These genes would now be inherited according to Mendel's laws. Evans modified the ESCs genetically by using retroviruses, which integrate their genes into the chromosomes. He successfully transfered such retroviral DNA from ESCs, through mosaic mice, into the mouse germ line. Thus, Evans used the ESCs to generate mice that carried new genetic material.

Capecchi and Smithies had demonstrated that genes could be targeted by homologous recombination in cultured cells. Evans had contributed in developing necessary vehicle to the mouse germ line–the ESCs. The next step was to combine the two. For their initial experiments both Smithies and Capecchi chose a gene (hprt) that was easily identified. This gene is involved in a rare inherited human disease (Lesch-Nyhan syndrome). Capecchi refined the strategies for targeting genes and developed a new method (positive-negative selection, see Figure 5) that could be generally applied.

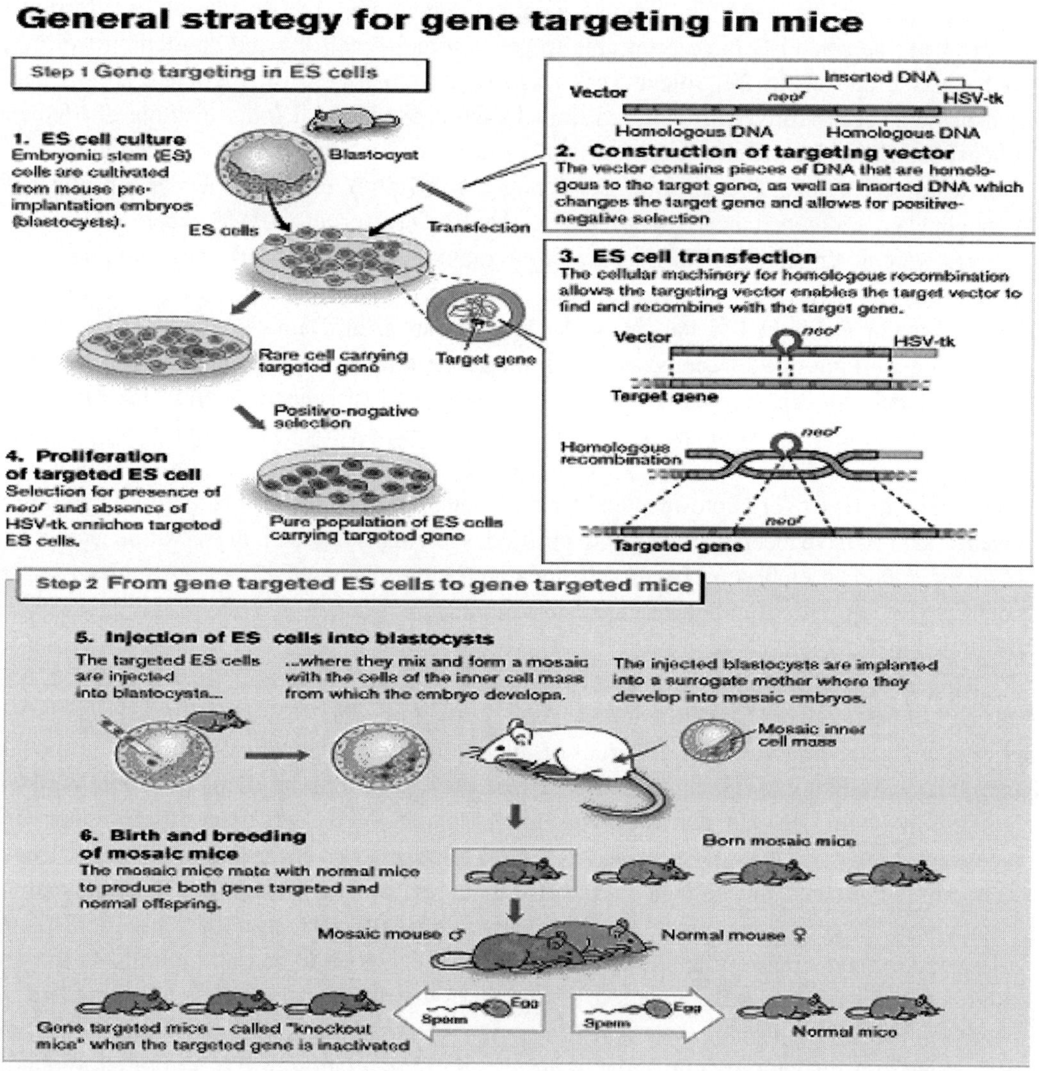

Reproduced from website: http://nobelprize.org/nobelprizes/medicine/laureates/2007/press.html

Figure 5. Gene targeting strategies in mice.

After the publication of the first report on homologous recombination in ESCs to generate gene-targeted mice in 1989, the number of reported knockout mouse strains has risen exponentially. It is now possible to introduce mutations that can be activated at specific time points, or in specific cells or organs, both during development and in the adult animal. Gene targeting has helped us in understanding the roles of many hundreds of genes in mammalian fetal development. The research work of Capecchis uncovered the roles of genes involved in mammalian organ development and in the establishment of the body plan and exlained the causes of several human inborn malformations. Applying gene targeting, Evans and Smithies developed several mouse models for the inherited human disease such as cystic fibrosis. Using these models he studied disease mechanisms and tested the effects of gene therapy for cystic fibrosis and the blood disease, thalassemia and other common human diseases such as hypertension and atherosclerosis.

The stem cell research in India is gaining momentum. At LV Prasad Eye Institute-Hyderabad, the scientists have been able to generate a 1-2 cell layered epithelium from the limbal tissue and about 20 patients diagnosed with limbal stem cell deficiency have been transplanted successfully with cultured limbal stem cells. The All India Institute of Medical Sciences (AIIMS)-New Delhi has successfully given stem cell treatment by 'injection method' to its cardiac patients. The institute has conducted the study under the leadership of Dr. P Venugopal, a renowned heart surgeon and the Ex-Director of AIIMS where 35 cardiac patients were given stem cell therapy and were monitored at 6, 12, and 18-months intervals. In the private sector, Reliance Life Sciences (RLS)-NaviMumbai, India backed by the industry major Reliance Ltd has taken the lead. It ranks third among the top 10 institutes worldwide working on stem cells, as stated in an NIH report. Another player in this arena is Manipal AcuNova, a collaborative venture between Manipal Education and Medical Group International India as well as AcuNova Life Sciences-USA. A technology has been established at Christian Medical College (CMC), Vellore for collection, isolation and purification of HSCs for haplo-identical haematopoietic stem cell transplantation. In addition, banana lectins have been isolated and purified showing stem cell preservation activities. Indigenous human embryonic stem cell lines are being generated at few institutions in the country [13]. A list of other research institutions associated with stem cell research is illustrated in Table 2.

There are several limitations to the therapeutic application of stem cells. For instance, it is difficult to harvest the ASCs from human tissues due to the fact that the stem cell population is very limited and harvesting techniques of internal organs are technically difficult and risky for adult stem cell donors excepting for bone marrow and cord blood stem cell. There is also a problem of immunological reaction between transplanted cells and tissues from donor and receipient. Moreover, the quality of an adult stem cell decreases according to the donor's age. This causes limitation on the concept of self-adult stem cell transplantation for degenerative disorders in target populations, which usually are senile. Because of these constraints associated to the therapeutic applications of ASCs, ESCs prove to be more promising as they exhibit enough flexible differentiation potential. However, by using ESCs, the problem of poor yield of cell number in case of ASCs can be overcome but ESCs also have got certain limitations. It is difficult to control the division and differentiation of ESCs in order to obtain the specific type of cells. ESCs can also cause problem of immune rejection. The therapeutic cloning strategy used to produce ESCs faces severe criticism of its relatedness to human cloning. Therefore these technical limitations and ethical controversies require an appropriate solution before furthering the application of stem cells in medicine. The advantages and disadvantages of using the adult and embryonic stem cells have been briefly described in Table 3.

Table 2. Institutions in India associated with stem cells research and supported by Department of Biotechnology (DBT), Government of India

Type of Stem Cell Research Facility	Institutions
CMC-DBT Centre for Stem Cell Research (cGMP facility)	Christian Medical College, Vellore
Network programmes	CMC, Vellore; SGPGIMS, Lucknow; PGIMER, Chandigarh; All India Institute of Medical Sciences (AIIMS), New Delhi; R&R New Delhi; National Centre for Cell Science (NCCS), Pune and National Brain Research Centre (NBRC), Manesar
Stem cell research facilities	PGIMER, Chandigarh; SGPGIMS, Lucknow; KEM Hospital, Mumbai and LVPEI, Hyderabad
Embryonic stem cell research	National Institute for Research in Reproductive Health (NIRRH), Mumbai; National Centre for Biological Sciences (NCBS), Bangalore; NCCS, Pune; NBRC, Manesar and Jawaharlal Nehru Centre for Advanced Scientific Research (JNCASR), Bangalore
Limbal stem cells	LVPEI, Hyderabad; R. P. Centre, AIIMS, New Delhi and Regional Institute of Ophthalmology, Kolkata
Neural stem cells	NBRC, Manesar; National Institute of Mental Health and Neurosciences (NIMHANS), Bangalore and NCCS, Pune and University of Hyderabad, Hyderabad
Mesenchymal stem cells	CMC, Vellore; SGPGIMS, Lucknow and Manipal Hospital, Bangalore
Pancreatic progenitor cells	National Institute of Nutrition (NIN), Hyderabad NCCS, Pune
Cardiac stem cells	Sree Chitra Tirunal Institute for Medical Sciences & Technology (SCTIMST), Thiruvananthapuram and AIIMS, New Delhi
Cancer stem cells	Indian Institute of Science (IISc), Bangalore

Table 3. Advantages and disadvantages of embryonic (ESCs) and adult stem cells (ASCs)

Embryonic stem cells (ESC)	Adult stem cells (ASC)
Advantages 1. More Flexible i.e. they have the potential to make any body cell. 2. Immortal i.e. one cell line can potentially supply endless amounts of cells with carefully defined characteristics. 3. Easily available: human embryos can be obtained from fertility clinics.	**Advantages** 1. Already somewhat specialized: Inducement may be simpler. 2. Immunocompetent i.e. recipients who receive the products of their own stem cells will not experience immune rejection. 3. Although less flexible but still may be: Adult stem cells may be used to form other tissue types. 4. ASCs have got mixed degree of availability i.e. Some adult stem cells are easy to harvest and others, such as neural (brain) stem cells, can be dangerous to the donor.
Disadvantages 1. Difficult to control i.e. the method for inducing the cell type needed to treat a particular disease must be defined and optimized. 2. At odds with a patient's immune system: It is possible that transplanted cells would differ in their immune profile from that of the recipient and so would be rejected.	**Disadvantages** 1. Minimal quantity: They are difficult to obtain in large quantities. 2. Finite: They don't live as long in a culture as embryonic stem cells. 3. Genetically unsuitable: The harvested stem cells may carry genetic mutations for disease or become defective during experimentation.

Stem cell research as it may seem at first sight is not so easy. Apart from many technical hurdles to be cleared even more difficult to address is the ethical issues, among which most serious are the allegations towards its similarity to human cloning. Although, so called therapeutic cloning strategy used in stem cell engineering is entirely different from human cloning, still the pro-life activists and different religious groups claim that stem cell research will encourage the exploitation of embryos for extraction of ES cells and vice-versa. The worldwide concerns of stem cell research is evident from the statement of Pope John Paul II in an address before International Congress on Transplants on 2000-AUG-29 in which he had said *"...[medical] methods that fail to respect the dignity and value of the person must always be avoided. I am thinking in particular of attempts at human cloning with a view to obtaining organs for transplants: these techniques, in so far as they involve the manipulation and destruction of human embryos, are not morally acceptable, even when their proposed goal is good in itself"*. United Nations General Assembly has banned all types of cloning including therapeutic cloning in a recently passed resolution. However, India is one of the few countries where the government is largely supportive of research in this segment.

Conclusion

Despite the wonderful therapeutic applications of stem cells in varied life threatening diseases, current inventions in stem cell research have raised several clinical / socio-ethical questions which need to be addressed prior to its real impact for the benefits of human beings could be realized. For example the answer to a question about the mechanisms that allow human embryonic stem cells and embryonic germ cells to proliferate *in vitro* without differentiating is not clearly known. Similarly, we need to understand about the genetic programming status of stem cells and the intrinsic controls that keep stem cells from differentiating and that direct the stem cells along a particular differentiation pathway to form one specialized cell over another, the universality of stem cells and factors responsible for stem cells to "home" to sites of injury or damage, the exact source, genesis, kind and location (specific tissue) of ASCs in the body (ASCs is known to remain in an undifferentiated state when all the cells around them have differentiated), the possibility to manipulate ASCs to increase their ability to proliferate in a culture plate so as to use ASCs as a suitable source of tissue for transplantation, the cellular and molecular signals that are important in activating a human pluripotent stem cell to begin differentiating into a specialized cell type, the stage of differentiation of stem cells best for transplantation (which might differ on a case by case) and screening drugs or toxins or for delivering potentially specific therapeutic drugs, the signals that regulate the proliferation and differentiation of stem cells that demonstrate plasticity, the passage of all pluripotent stem cells through a progenitor / precursor cell stage while becoming specialized, the possibility of maintaining the progenitor / precursor cell as optimal cells for therapeutic transplantation etc. However, the research into potential alternatives for therapeutic cloning should be stimulated. The widespread applications that stem cell research offers in the therapeutic solutions to most of the debilitating diseases make it imperative to explore different dimensions of this area, which could be utilized primarily for human welfare.

References

[1] Becker AJ, McCulloch EA & Till JE (1963) .Cytological demonstration of the clonal nature of spleen colonies derived from transplanted mouse marrow cells. *Nature,* 1963,197, 452–454.

[2] Siminovitch L, McCulloch EA & Till JE (1963). The distribution of colony-forming cells among spleen colonies. *Journal of Cellular and Comparative Physiolog,* 62, 327–336

[3] A representative stem cell, http://www.rte.ie/news/2007/1008/nobel.html

[4] Cumano A, Ferraz JC, Klaine M, Di Santo JP & Godin I. (2001) Intraembryonic, but not yolk sac hematopoietic precursors, isolated before circulation, provide long-term multilineage reconstitution. *Immunity,*15(3),477-485.

[5] Chen LB, Jiang XB & Yang L (2004). Differentiation of rat marrow mesenchymal stem cells into pancreatic islet beta-cells. *World Journal of gastroenterology*,10 (20), 3016–3020.

[6] Netter & Frank H.(1987). *Musculoskeletal system: anatomy, physiology, and metabolic disorders*. Summit, New Jersey: Ciba-Geigy Corporation ISBN 0914168886, p.134.

[7] Brighton, Carl T. & Hunt RM (1991).Early histologic and ultrastructural changes in medullary fracture callus. *Journal of Bone and Joint Surgery*, 73-A (6), 832-847.

[8] Engler AJ, Sen S, Sweeny HL & Discher DE (2006). Matrix Elasticity Directs Stem Cell Lineage Specification. *Cell*, 126 (4), 677–689.

[9] Ryan JM, Barry FP, Murphy JM & Mahon BP (2005).Mesenchymal stem cells avoid allogeneic rejection. *Journal of Inflammation* (Lond), 2, 8.

[10] Rubin R & Strayer DS (2007). *Rubin's Pathology: Clinicopathologic Foundations of Medicine*. Lippincott Williams & Wilkins. p.90.

[11] Gage FH (2000).Mammalian neural stem cells. *Science*, 287(5457), 1433-1438.

[12] Work of Nobel Laureates:(http://nobelprize.org/nobel_prizes/medicine/laureates/2007/press.html).

[13] Dureha R (2006). The promise of stem cells technology in India, *Biospectrum*, Feb 09.

[14] Thomson JA, Itskovitz-Eldor J, Shapiro SS, Waknitz MA, Swiergiel JJ, Marshall VS & Jones JM (1998). Embryonic stem cell lines derived from human blastocysts. *Science* (New York), 282 (5391), 1145–1147.

[15] Cibelli JB, Lanza RP, West MD & Ezzell C (2001). The first human cloned embryo. *Scientific American*. http://www.sciam.com/article.cfm?id=the-first-human-cloned-em

[16] Shostak S (2006). (Re)defining stem cells. *Bioessays*, 28 (3), 301–308.

[17] Takahashi K & Yamanaka S (2006). Induction of pluripotent stem cells from mouse embryonic and adult fibroblast cultures by defined factors.*Cell*,126 (4),663–676.

[18] De Coppi P, Bartsch G, Siddiqui MM, Xu T, Santos CC, Perin L, Mostoslavsky G, Serre AC, Snyder EY, Yoo JJ, Furth ME, Soker S, Atala A (2007).Isolation of amniotic stem cell lines with potential for therapy. *Nature Biotechnology*, 25 (1), 100–106.

[19] Cyranoski D (2007).Simple switch turns cells embryonic. *Nature*, 447(7145), 618–619.

[20] Mitalipov SM, Zhou Q, Byrne JA, Ji WZ, Norgren RB & Wolf DP (2007). Reprogramming following somatic cell nuclear transfer in primates is dependent upon nuclear remodeling. *Human Reproduction*, 22 (8), 2232–2242

[21] Takahashi K, Tanabe K, Ohnuki M, Narita M, Ichisaka T, Tomoda K, Yamanaka S (2007). Induction of pluripotent stem cells from adult human fibroblasts by defined factors. *Cell*, 131(5), 861–872.

[22] Yu J, Vodyanik MA, Smuga-Otto K, Antosiewicz-Bourget J, Frane JL, Tian S, Nie J, Jonsdottir GA, Ruotti V, Stewart R, Slukvin II & Thomson JA (2007). Induced pluripotent stem cell lines derived from human somatic cells. *Science*, 318(5858), 1917–1920.

[23] Chung Y, Klimanskaya I, Becker S, Li T, Marc Maserati M, Lu S-J, Zdravkovic T, Ilic D, Genbacev O, Susan Fisher, Krtolica A & Lanza R (2008). Human embryonic stem cell lines generated without embryo destruction. *Cell Stem Cell*, 2,113-117.

[24] French AJ, Adams CA, Anderson LS, Kitchen JR, Hughes MR & Wood SH (2008). Development of human cloned blastocysts following somatic cell nuclear transfer (SCNT) with adult fibroblasts. *Stem Cells Express*, 26, 485.

[25] Centeno CJ, Busse D, Kisiday J, Keohan C, Freeman M & Karli D (2008). Increased knee cartilage volume in degenerative joint disease using percutaneously implanted, autologous mesenchymal stem cells. *Pain Physician,* 11 (3), 343–353.

[26] Conrad S, Renninger M, Hennenlotter J, Wiesner T, Just L, Bonin M, Aicher W, Bühring HJ, Mattheus U, Mack A, Wagner HJ, Minger S, Matzkies M, Reppel M, Hescheler J, Sievert KD, Stenzl A, Skutella T (2008). Generation of pluripotent stem cells from adult human testis. *Nature,* 456 (7220), 344–349.

[27] Baker M (2008).Embryonic-like stem cells from a single human hair. *Nature Reports Stem Cells,* doi:10.1038/stemcells.2008.142.

[28] Woltjen K, Michael IP, Mohseni P, Desai R, Mileikovsky M, Hämäläinen R, Cowling R, Wang W, Liu P, Gertsenstein M, Kaji K, Sung HK& Nagy A (2009). piggyBac transposition reprograms fibroblasts to induced pluripotent stem cells. *Nature*, 458, 766-768.

[29] Kaji K, Norrby K, Paca A, Mileikovsky M, Mohseni P & Woltjen K (2009). Virus-free induction of pluripotency and subsequent excision of reprogramming factors. *Nature,* 458, 771, dói:10.1038/nature07864.

In: Recent Trends in Biotechnology, Volume 1
Editors: M.P. Singh, A. Agrawal and B. Sharma
ISBN 978-1-60876-148-7

Chapter 9

ETHICS AND BIOTECHNOLOGY

Arun Kumar Singh[1] and Sarita Singh[2]
[1] Department of Philosophy, T.D. College
Jaunpur (U.P.) India
[2] Department of Humanities and Social Sciences, U.N.S. Institute of Engineering & Technology, V.B.S. Purvanchal University
Jaunpur – 222001 (U.P.) India

ABSTRACT

There are some common tenets to guide our actions and these are accepted as paradigm of morality in almost all human societies irrespective of differences in creed and culture. But in specific cases often unanimity is lacking. One basic criterion of morality is the acceptance of responsibility by the decision taker. New researches in the field of medical sciences are creating new problems which were hitherto not faced. New techniques in biotechnology have paved way for transplantation for kidney, heart and liver successfully. *In vitro* fertilization, surrogate mothership, researches in plelipotent stem cells which are capable of cloning are creating moral dilemma for the scientists. These techniques have immense capacity for the wellness of human beings. But there is every possibility of their misuse. Should researches in biotechnology be allowed unconditionally or there be some guiding principles? If the technique of cloning is perfected what will be the fate of the society which is based on conjugal love and trust? Should abortion and euthanasia be allowed? These questions need a clear answer. In our view no such laws can be framed which can guide scientists in each and every case. Each case is specific and requires particular attention. There should be some norms for individual decision taker which are given in spiritual book of Hindu 'Shrimad Bhagavata Geeta'. With the help of this technique of decision making specific cases can be dealt with.

Introduction

Ethics is concerned with the 'oughtness' of our actions. What should be done in a particular circumstance? What should be the aim of our actions? What should be the criteria of calling an action 'good' or 'bad', 'right' or 'wrong'? No clear cut answers to these questions are possible. Many philosophers have debated over these questions since long but they have not come to a unanimous resolution. Those views can not be called synonymous to each other. A thinker lives in a particular culture and milieu and his decisions are not totally immune to their influences. The difference between a professional of other field and a philosopher is that the latter is supposed to think disinterestedly and his answer may be helpful to solve the problem concerning ethical dilemma. But the entire onus lies on the decision maker who is directly involved in the case.

Bioethics and Social Issues

Bioethics is a new discipline and this word was first coined by a biologist S.V.R. Polter in 1970. This was the era when the physicists who explored the inner structure of matter and saw the destructive use of atomic energy and the biologist who unraveled the mystery of human life by mapping genome exposed their anguish and concern for the social implication of their knowledge. To deal with ethical implication of scientific findings Kennedy Institute of Ethics was founded at Georgetown University, America in 1971. *The Encyclopedia of Bioethics* which the institute began to plan in 1972, defined bioethics as 'the study of the ethical dimensions of medicine and biological Sciences' [3,5]. The code of conduct for medical practitioners is not a new phenomenon. All the great cultures - Chinese, Indian, Roman and Arabic where medical practice was in vogue some sort of ethical code was defined. It was generally accepted that the relation between patient and a physician is very pious and latter is a sort of guardian to the former. The physician will in no case divulge the secret of the patient and his attitude should be to cure him not to fleece him. In India it was imperative for the 'Vaidya' to charge the minimum for his services. But with the advance of science new technologies were developed and their application in medical sciences created problems which were not faced by the ancient followers of this discipline. Old ideas about medical morality were found inadequate to meet the challenge posed by new developments in Biotechnology.

Ethical Problems Concerning Medical Science

A brief description of some problems of Medical science is given below. Science is based on experimentation and medical science is concerned with human well being. New medicine can be prescribed only after its efficacy is established and for this it should be tested on human beings. But all experimentation is a journey into the unknown and they involve some risk. In Second World War Nazis used prisoners as object of experimentation and so many persons suffered heavily. Even in America some unscrupulous physicians experimented on black people causing great physical ailment to them. American Government established a

National Commission to protect the human subjects from biomedical research. The commission proposed three principles to guide the research. These are – respect for persons, beneficence and justice, which means informed consent, the assessment of risk in relation to benefit and equitable selection of subjects for research [1,4].

Transplantation of organ from one person to another is possible and this technique has benefited so many persons. Death from kidney failure is avoidable and kidney from a healthy person can be transplanted in the body of a person suffering from kidney failure but for this the physicians have to invade the body of healthy person which is akin to obstruct the natural function of the body. Sometimes poor people are enticed to sell their healthy organs and this advancement in medical technology gives impetus to unethical practice in the society.

In 1968 Dr. Christian Bernard successfully transplanted heart from a dying person to the other who lived for several months. Heart transplantation is different from kidney transplantation. In kidney transplantation one functioning live kidney is taken from a healthy body and the remaining one is sufficient for the donor for living a normal life. In heart transplantation a physiologically living organ must be taken from a person who is already legally dead. This requires a definition of death. The first problem of ethical concern is the time of death. When the person is supposed to be dead? When the brain is not functioning or when the heart is not palpitating. What is the distinction between life and death? Is sustained physiological activity through artificial means is life? When a person has lost the capacity of internal and external stimuli, can he still be called alive? Without finding a clear cut definition of death, how a functioning heart can be taken for transplantation?

Human reproduction was considered to be guided by some laws established by God or for the atheist by a process set by Nature. But with the advancement of scientific knowledge, this has also become an object of human manipulation. Now sperm and ovum taken from two persons – man and woman can be fertilized in the laboratory and the developed egg can be implanted in the womb of woman who can give birth to a child. If the mother is not giver of the ovum, she is called surrogate mother. Now the question of motherhood arises. Is the woman who donated the ovum or the woman who gave birth to the child be called the legal mother? Surrogate motherhood is creating new legal and ethical problems. Recently, a Japanese couple hired the services of surrogate mother in India. But before the birth of the child the couple got divorced. The legal problem is as to who should be given the custody of the child, the man who donated the sperm or the woman who gave the ovum. Such problems were not debated before.

In late twenties further advanced developments in biotechnology created new problems. In natural process of reproduction sperm from man and ovum from woman spliced to make a fertilized complete cell which develops as an embryo. In cloning female egg is spliced with a cell taken from the body and the fertilized cell is implanted in the female. The resulting offspring is identical with the donor of the skin cell. Using this technique a lamb named Dolly was produced. Scientist are able to isolate stem cells from human embryos, these cells are immature but plelipotent which can be grown into infinite variety of human cells. This technique has immense capacity which can be used for the benefit of human beings. Different human organs can be grown into laboratory and transplantation of kidney, heart, liver can be possible without taking the services of a donor. These cells can be used to create clone infants too and if these techniques become a success, any dictator can create havoc by forcing scientist to create an army of clone persons. The natural process of reproduction which involves men and women will be ignored. The pleasure and solace of conjugal life which is the very basis of culture and society will be in jeopardy. Should such researches be allowed? This is the question which is debated by theologiceans, philosophers and politicians all over the world. In 2001 the then president of United States of America George Bush announced that public funds could not be used for such research except in highly restricted circumstances. But recently President B.H. Obama has removed this restriction over stem cell research. A well defined ethical code is imperative to guide such experiments.

Apart from above mentioned problems which are due to the advancement in biotechnology, some old problems also lack unanimous approval by different cultural groups and societies. Can an unwanted embryo be terminated by abortion? Should a person suffering from incurable disease be given the option of choosing death to avoid excruciating suffering? But accepting abortion and euthanasia put us in a state of moral dilemma. Are we authorized to terminate the life of an unborn child? Is it not akin to murder? If euthanasia or mercy killing is allowed, will it not violate the very purpose of medical profession? A person suffering from cancer or AIDS will be neglected by family members so that he may opt for euthanasia. Will this not be against the lofty ideals of humanity?

Bioethical Dilemma and Solutions

Whenever an ethical dilemma arises whether it concerns a person or is consequential to society, a decision is to be taken by a person or by a group of persons. No traditional ethical theory is capable of guiding us in each and every case. We can take some advice in the art of decision making by the techniques of 'Shrimad Bhagavata Geeta' which is a divine pathfinder for the devotees since the time immemorial. Although the problem faced by the scientist is different from that faced by mythological Arjuna but the dilemma of indecision is same. 'Geeta' is the unique book, it is not only for seekers of liberation or mukti, its advice is very helpful for secular activities of our lives. One of the most fascinating lessons of 'Geeta' is a novel technique of decision making with the help of which Arjuna found his way out of an excruciating state of indecision. These techniques have been described in a very enlightening book on 'Geeta' entitiled '*The Gita – A workshop on expansion of self*' [6]. A brief description is given below which may be helpful to a scientist who is facing a difficult problem of continuing or discontinuing an experiment causing moral dilemma to him. The techniques of decision making has not been outlined explicitly on any place in the text of 'Geeta' but can be deduced from the teaching spread in 18 chapters. It has actually been followed and practiced by the true adherent of 'Geeta'. It is a technique based on the objective assessment of the situation as well as the psycho – sociological make up of the decision taker. Outlined below are the different steps of this technique.

The decision maker should explore all the possibilities without any bias or prejudice to any one of the possibilities. As a result of dispassionate reasoning he may arrive at a certain decision. If it is not possible, he should take the advice of an enlightened mentor and if the mentor is incapable of by beaconing him a path to tread upon, the third step should be taken. This is to de ego one's outlook. One should disinterestedly analyze the motive of one's action. He should find out whether the desire to achieve a result is needed to satisfy one's own self esteem or really the motive is to gain knowledge to further the philanthropic cause. It is quite possible that the very result desired by the decision taker is consciously or unconsciously guided by the desire to bolster his or her ego. Generally, most of our decisions are guided by consideration of ego and consciously we ascribe lofty ideals to justify them. That is why the step to de ego our motive is necessary. If the state of indecision still persists, **he should take the fourth step which is to define one's role in decision making.** Whether a person who is involved in the act is simply an instrument doing his job without any personal involvement or a rational responsible person whose act is going to influence the society. He should take himself as a person who is setting an example for the others to emulate him and be prepared to take the responsibility of the consequences of his action.

The first four steps are meant for the preparation of the individual decision maker. The next four steps are designed to test the options before them, this is test of rationality. The criterion of rationality is universality. The decision should be such that it can be applied in all such cases irrespective of social or cultural differences. The sixth step should be to be careful about the dignity of human being. Each human being is divine in nature and none is authorized to make other his or her instrument. Each individual is an end in itself. While taking some moral decision we can adopt a course of action which can be passed by everyone without detriment to any one else. Any criterion for testing the correctness of a course of action should be based on the standard of rectitude chosen by the decision taker. The next criterion of decision making is self identification. One should be happy in taking the decision and feeling of remorse should not be associated with it. In the final step one should visualize the final result of the action. If decision taker feels that his personality or image is not damaged and he gets the admiration and applause for his action, he should compliment himself for reaching a correct decision.

CONCLUSION

In nutshell we can say that while taking the decision which is going to affect others, we must be free from personal prejudice, cultural bias and be careful about the implication of decision for humanity at large. One should not violate the lofty ideals cherished by all the civilized society and decision should be based on the standard of rectitude. Human dignity and well being should be our aim. Taking into consideration the normal of decision making specific laws can be framed to promote the technical advancement in the field of medical sciences which may not be detrimental to human dignity. But these laws should be resilient and strict application will go against the very purpose of their framing.

REFERENCES

[1] Childress, J., Meslin, E. and Shapiro, H. (eds.) 2005. *Belmont Revisited. Ethical Principles for Research with Human Subjects.* Gorgetown University Press, Wshington, D.C.

[2] Fox, R. and Swazey, J.1974. *The courage to fail. A Social View of Organ Transplant and Dialysis.* University of Chicago Press, Chicago.

[3] Jonsen, A.R.2007. A history of bioethics as discipline and discourse. In: N.S. Jecker, A.R. Jonsen and R.A. Pearlman (Eds.), *Bioethics – An Introduction to the History, Methods and Practice*. Jones and Bartlett Publishers, London, pp. 3-16.

[4] Jonsen, A., Veatch, R. and Walters, L. (Eds.) 1998. *Source Book of Bioethics.* Gorgetown University Press, Wshington, D.C.

[5] Reich, W. 1978. Introduction. In: W. Reich (ed.), *The Encyclopedia of Bioethics* vol. I. The Free Press, New York, pp xix – xx.

[6] Singh, I.P.1977. The Gita in everyday life (chapter 7). In: *The Gita: a Workshop on the Expansion of Self.* Somaiya Publication Pvt. Ltd, Bombay, pp. 130-145.

In: Recent Trends in Biotechnology, Volume 1
Editors: M.P. Singh, A. Agrawal and B. Sharma
ISBN 978-1-60876-148-7

Chapter 10

IN VIVO INDUCED ANTIGEN TECHNOLOGY

Brahm S. Srivastava and Ranjana Srivastava
Microbiology Division, Central Drug Research Institute, Lucknow

ABSTRACT

IVIAT, abbreviated for *in vivo* induced antigen technology, is one of the recently invented techniques for identification of genes of a disease causing pathogen specifically induced upon infection of host. In simpler words these genes are switched on *in vivo* but remain switched off while growing *in vitro* outside the host. Since these genes are transcribed in the host environment, it is assumed that they may be essential for infection process, survival in the host environment and progression of disease. Identification of *in vivo* specific genes may thus be important in development of chemotherapeutics, vaccines and diagnostics. IVIAT offers several advantages over other techniques developed to select out genes induced *in vivo*. IVIAT circumvents the use of animal models and can be applied to detect temporal expression of antigens through infection process.

INTRODUCTION

Most of the human pathogens, such as bacteria, protozoa and helminths complete their life cycle in at least two environments; one, within the human host transiently or for long duration and the second, in another animal or stay as free living. Hence the pathogen has to adapt to respective environments for survival and in doing so it has been found that the pathogens express various proteins which are essential for survival and progression of disease. These proteins are turned on in one condition and off in another environment. It is generally accepted that the antigens expressed in the human host may contribute to the pathogenicity of infecting organism. In fact these antigens are effectively virulence determinants, thus may be important in development of therapeutics, vaccines and diagnostics.

Realizing the importance of *in vivo* induced antigens, several techniques have been developed to specifically identify these antigens. To name a few, these techniques are called *in vivo* expression technology [1], signature-tagged mutagenesis [2, 3], differential fluorescence induction [4], selective capture of transcribed sequences [5] and *in vivo* induced antigen technology [6, 7]. Although they serve the same objective, they differ in concept and logistics but are based on genomic approach. They select out *in vivo* induced genes.

IVET, *in vivo* expression technology, is the first technique developed to identify *in vivo* induced genes of bacteria [1]. It is a promoter trap strategy in which in a suitable plasmid vector, random DNA fragments are cloned in frame upstream of a reporter gene. In the original scheme, the reporter gene was a promoterless gene encoding for an essential growth factor of bacteria. The library of the random DNA fragments immediately upstream of promoterless reporter gene were injected in mice. Only those promoters switched on *in vivo* would support synthesis of the essential growth factor and thus the survival of bacteria. Nucleotide sequence of the cloned DNA driving the expression leads to the identification of *in vivo* induced gene. The strategy offers positive selection. A number of other reporter genes have been used in IVET.

Signature tagged mutagenesis (STM) utilizes a library of sequence (acting as signature) tagged mutant bacteria to infect an animal model and those not surviving are considered as mutated in the gene required for growth *in vivo*. In this way as compared to IVET, STM is a negative selection technique but has been used to identify *in vivo* induced genes [2, 3].

In differential fluorescence induction technique, promoterless green fluorescent gene (GFP) is the reporter gene upstream of which a promoter library is cloned. Bacteria expressing GFP are segregated from those lacking GFP expression *in vivo* [4].

It should be stressed that these strategies require an animal model mimicking natural human infection. IVIAT technique, however, overcomes the need of an animal model. The concept of IVIAT was published in year 2000 [6] and by 2002 onwards results on application of IVIAT in identification of *in vivo* induced genes of microbial pathogens had began to be published [7-11]

IVIAT is based on the premise that antibodies to the antigens of pathogen of interest are produced when the immune system of the host encounter the pathogen. Hence, antibodies present in the clinical sera can be used to identify antigens of the pathogen which are elaborated in the host. Since the serum may have antibodies to all the antigens of bacteria, adsorption of serum with antigens of the pathogen grown *in vitro* would remove antibodies to *in vitro* specific antigens and leave in the serum the antibodies produced in response to antigens expressed only *in vivo*. If absorbed serum with remaining antibodies are used to screen a protein expression library of the same pathogen, antigen-antibody reaction can lead to identification of genes encoding the antigens during *in vivo* growth (Figure 1).

IVIAT is a simple and convenient method to identify array of genes transcribed following infection. Essentially, it requires a genomic protein expression library of the pathogen of interest, an *in vitro* grown culture of the cognate organism and a pool of clinical sera obtained from patients infected by the same pathogen. The method can be accomplished in few steps (Figure 1 and Table 1).

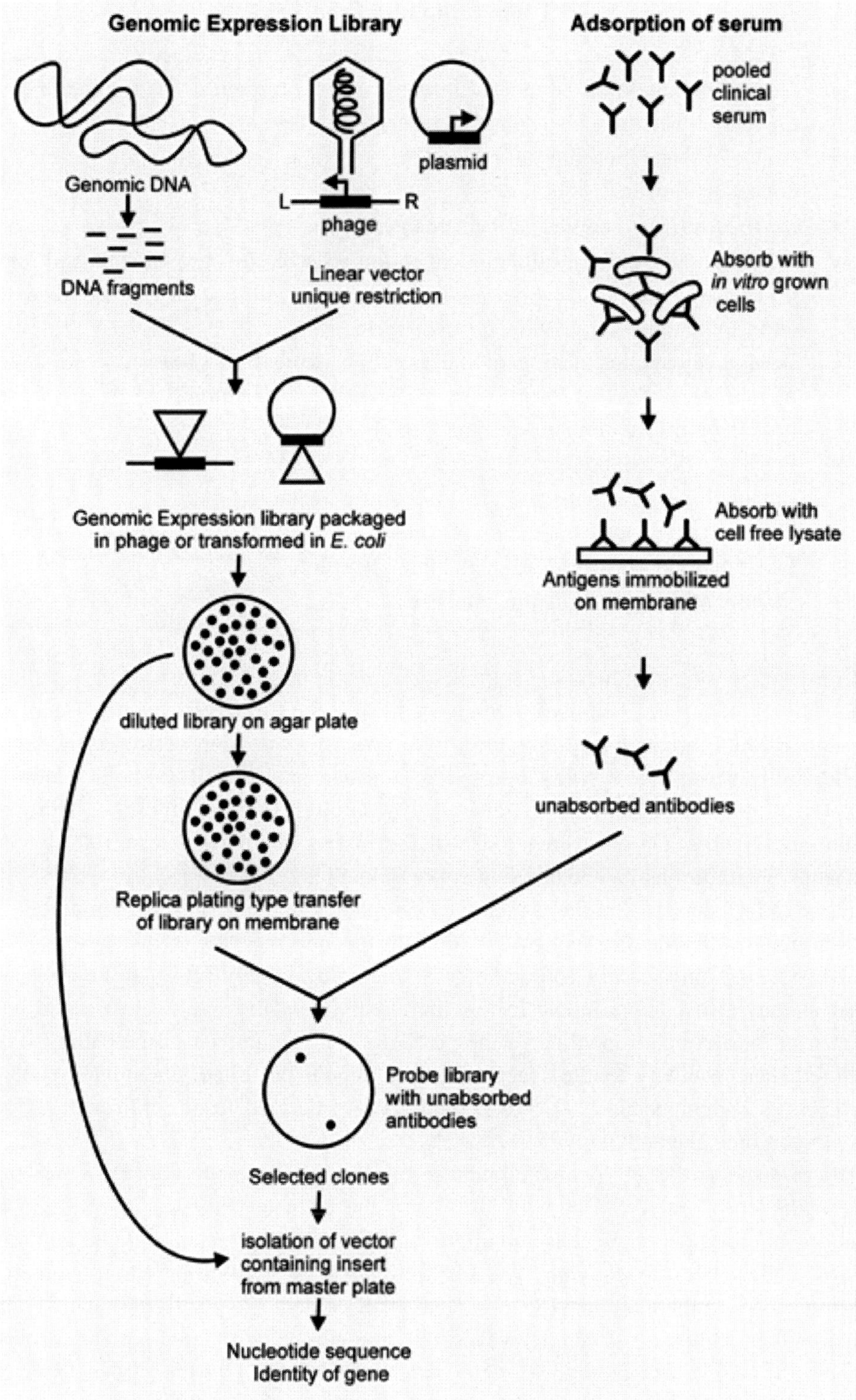

Figure 1. IVIAT in search of genes expressed uniquely in human host.

Table 1. Steps in IVIAT

- Genomic expression library of the pathogen of interest in a suitable protein expression vector, such as plasmid or bacteriophage λ.
- Collection and probing of clinical sera obtained from patients infected with cognate pathogen.
- *in vitro* cultivated culture of the same pathogen.
- Thorough adsorption of the clinical serum with *in vitro* grown cells and removal by centrifugation of bacterial antibody complex and recovery of absorbed serum and another adsorption with cell free lysate.
- Seeding by replica plating of the genomic expression library on nitrocellulose membrane, induction of protein expression and immobilization of expressed protein on the membrane.
- Probing the expressed protein by absorbed serum using appropriate immunological reagents and identification of reactive clones.
- Selection of clones from the master plate.
- Isolation of DNA insert and nucleotide sequence.
- Identity of the gene from the database.

Construction of an inducible genomic protein expression library of the pathogen of interest is very important. The library would be ideal if the full complement of genes may be represented. From the published literature, two types of expression vectors have been used; plasmid and bacteriophage λ with isopropyl-B-D-thiogalactoside (IPTG) inducible promoter. Fragments of genomic DNA may be generated by partial restriction digestion, hydroshearing or blunt-end cloning. On an average, size of the inserts ranging between 1000 to 3500 bp (base pair) in a library may yield efficient expression.

Since IVIAT depends on the presence of antibodies in the serum, it is important to pay attention while collecting the sera from patients, especially keeping in mind the goals set by the investigator. Convalescent and acute sera containing largely IgG may be collected and pooled. It may also be appropriate to test the reactivity of the sera against crude lysate of pathogen of interest before pooling. If the profile of a given antigen of the pathogen is to be monitored then reactivity to the specific antigen should be tested. Creating a pool of 5-8 individuals is recommended [11]. We pooled 25 sera derived from confirmed TB patients before undergoing chemotherapy [7].

IVIAT requires that the pooled clinical serum is pre-adsorbed with *in vitro* grown cells of the pathogen of interest. One way is to grow the cognate pathogen in optimum growth conditions in broth, pellet the cells and after discarding the supernatant, diluted pooled serum is added to the pellet of intact cells. The adsorption is allowed to occur. In principle, it is like antigen-antibody agglutination. The complex is discarded and the supernatant containing unabsorbed antibodies are stored. However, this may not be considered perfect adsorption as antibodies to internal or membrane proteins synthesized *in vitro* have not been removed. It is highly recommended to absorb the serum with cell lysates of *in vitro* grown organism. After growing cells *in vitro*, a cell free lysate is prepared by breaking cells by sonication and the antigens in the lysate are bound to nitrocellulose membrane. After blocking empty spaces on the membrane, the diluted serum is incubated with antigen bound membrane. A detailed account of the procedure of binding the antigens to nitrocellulose membrane, blocking the membrane and probing with antibodies has been published [12, 13]. This removes antibodies to all the antigens present in cells grown *in vitro*. The absorbed serum is now ready to probe expressed proteins from the genomic expression library. However, the serum may have cross reacting antibodies to *Escherichia coli* and to bacteriophage λ, which might give false positive signals while screening of expressed proteins in the genomic library. Therefore adsoption of the serum with antigens of *E. coli* and bacteriophage λ is also recommended.

The genomic expression library can now be screened with absorbed serum. The absorbed serum contains antibodies to those antigens which were expressed during *in vivo* growth in the human host. If the library was generated in an inducible plasmid expression vector and transformed in *E. coli,* the library will be appropriately diluted and plated on nutrient agar plates containing the plasmid selection antibiotic and IPTG. Care should be taken to avoid confluent growth. Each plate is replica blotted on nitrocellulose membrane after small colonies visible to naked eye have grown. The blotted membrane is placed on another nutrient agar plate and the colonies are allowed to grow to small visible size. The colonies are then treated with chloroform-SDS gently to allow lysis to occur and expressed proteins are released and bound to nitrocellulose membrane. The choice of *E. coli* strain is important. Engineered *E. coli* strains such as BL21 (DE3) are commercially available. In case of λ expression system, the phage library amplified in *E. coli* strain Y1088 is suitably diluted. The library is screened by plating approximately 10^3 plaque-forming units with *E. coli* strain Y1090 on 90 mm plate and incubated for 3 hours at 42^0C and wait for small phage plaques to appear. Care should be taken to avoid confluent lysis. Nitrocellulose membrane of the size of the agar presoaked in IPTG is placed over the agar surface to allow transfer. The plate with over laid membrane remains incubated for 2-3 hours at 37^0C (for finer details [12]).

After immobilization of the expressed proteins on nitrocellulose membrane, identification of clones expressing immunoreactive proteins are identified by an approach which is essentially ELISA. The membrane is treated with primary antibodies (absorbed sera) followed by secondary antibody commercially available and then developed with suitable substrate. The positive clones (be in bacteria or phage) should be confirmed along with appropriate positive and negative controls by another round of blotting and antibody mediated screening. The positive clones must be purified to ensure single insert expressing the reactive protein.

The insert DNA needs to be characterized. The expression vector (plasmid or phage) with insert DNA (of the pathogen of interest) is isolated and sequenced. If the sequence of the genome of the pathogen is known, the identity of the gene can be easily found out. IVIAT leads to identity of a gene which may have an important role in the disease process. The function of these genes may be known or unknown which can be determined by classical approaches including mutagenesis, knock-out or over-expression.

IVIAT is a comparatively new technique but it has been successfully applied in searching virulence factors of *Mycobacterium tuberculosis* [7], *Vibrio cholerae* [8], *Vibrio vulnificus* [9], and *Candida albicans* [10]. The list of genes encoding the virulence factors of these organisms is really impressive. Already, extensive work has been done on elucidation of virulence factors of *V. cholerae* and it is encouraging to find that IVIAT has picked up many of these known genes encoding the virulence factors [8]. So are the results with *C. albicans* [10]. *M. tuberculosis* is a complex pathogen and search for virulence factors has only recently begun. Several genes of *M. tuberculosis* with putative role in virulence were selected by IVIAT [7]. Later analysis revealed that one of the genes Rv3303c has an important function in pathogenesis by protecting mycobacteria against oxidative burst [14]. Therefore, application of IVIAT in discovering virulence factors and the genes involved in progression of disease appears highly encouraging.

It will be pertinent to compare IVIAT and IVET as the latter technique has also yielded useful information efficiently and has been used since 1993. Originally discovered by Mahan *et al.* [1] the technique offers to identify genes which are transcribed *in vivo*. They developed IVET using *Salmonella enterica* serovar *Typhimurium*. The bacterial strain was a purine auxotroph (Δ*pur*A) which could not grow *in vitro* or in mouse unless purine was supplemented *in vitro* in growth medium or complemented in the latter. A beautiful strategy was devised to complement *pur*A (Figure 2). A promoterless purine gene was engineered on plasmid upstream of which random genomic fragments of *S. enterica* serovar *Typhimurium* were cloned. This library was transformed in *Salmonella* Δ*pur*A and infected in mice. Bacteria will survive in mice only when purine auxotrophy was complemented. The promoterless purine gene on the plasmid will become functional only after being transcribed by an inducible promoter cloned upstream. The identity of *in vivo* induced gene is revealed by nucleotide sequence. Modifications of IVET have been made with respect to the reporter or selection genes and the technique has been successfully applied to other organisms [15]. IVET has also been used by us to identity genes of *M. tuberculosis* expressed *ex vivo* in macrophage cell line, in the lungs of mice and during anaerobic persistence [16-18]. A promoter-trap vector with an artificial bicistronic operon containing promoterless green fluorescent protein (GFP) and kanamycin resistant (kan^r) genes was used to identify the pool of genes upregulated *in vivo* (Figure 3). Most of these genes were associated with virulence.

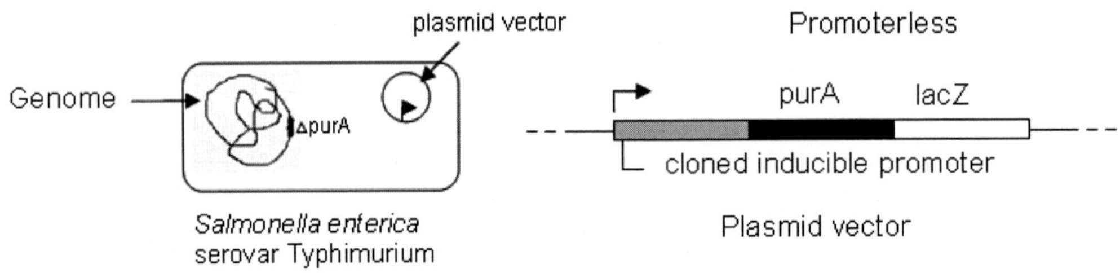

(Adapted from Mahan *et al.*[1], scale arbitrary)

Figure 2. IVET; selection by complementation of purA.

In sharp contrast to IVET, IVIAT does not require the use of an animal model to identify *in vivo* induced genes. It takes clinical patient sera which logically contain the full repertoire of antibodies to all the antigens, both *in vitro* and *in vivo* expressed, of the pathogen. The sera would be absorbed with antigens expressed *in vitro* before actually being used to identify *in vivo* induced genes. The genomic protein expression library is probed with absorbed sera.

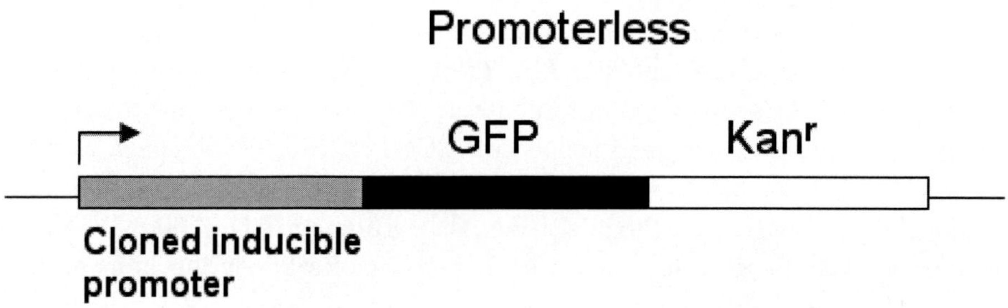

(Adapted from Srivastava *et al.* [16], scale arbitrary)

Figure 3. IVET; selection based on fluorescence and kanamycin resistance.

CONCLUSION

IVIAT technique has several advantages over other techniques used for similar purpose. The most striking advantage is that IVIAT is not dependent on the use of animal model. It is an antibody based selection technique. These antibodies have been produced in the natural host in response to infection of the pathogen of interest. Animal models do not necessarily mimic the host environment and immune responses. IVIAT provides opportunity to use sera of various stages of infection and can be used to map the temporal expression of virulence factors in the host. Variability in individual immune responses can be avoided since a pooled patient serum is used in screening. Genomic protein expression library from clinical strains of the same pathogen can be pooled to increase the repertoire of antigens. So far, IVIAT has been optimized to select genes responding to humoral immune response and there is no report of selection of those involved in cell mediated immune responses. In conclusion, IVIAT has proved to be a very potent technique to select genes induced uniquely *in vivo*. These genes are likely to be virulence factors required for survival and pathogenesis. Although the technique is new, yet since 2002 IVIAT has been applied to explore virulence factors of invasive and noninvasive bacterial and fungal pathogens.

REFERENCES

[1] Mahan, M.J., Slauch, J.M. & Mekalanos, J.J. (1993) Selection of bacterial virulence genes that are specifically induced in host tissues. *Science*, 259, 686-688.

[2] Hensel, M., Shea, J.E., Gleeson, C., Jones, M.D., Dalton, E. & Holden, D.W. (1995) Simultaneous identification of bacterial virulence genes by negative selection. *Science*, 269, 400-403.

[3] Camacho, L.R., Ensergneix, D., Perez, E., Gicquel, B & Guilhot, C. (1999) Identification of a virulence gene cluster of *Mycobacterium tuberculosis* by signature-tagged transposon mutagenesis. *Mol. Microbiol.* 34, 257-267.

[4] Valdivia, R.H. & Falkow, S. (1997) Fluorescence-based isolation of bacterial genes expressed within host cells. *Science*, 277, 2007-2011.

[5] Graham, J.E. & Clark-Curtiss, J.E. (1999) Identification of *Mycobacterium tuberculosis* RNAs synthesized in response to phagocytosis by human macrophages by selective capture of transcribed sequences (SCOTS). *Proc. Natl. Acad. Sci. USA* 96, 11554-11559.

[6] Handfield, M., Brady, L.J., Progulske-Fox, A. & Hillmam, J.D. (2000) IVIAT: a novel method to identify microbial genes expressed specifically during human infections. *Trends Microbiol* 8, 336-339.

[7] Deb, D.K., Dahiya, P., Srivastava, K.K., Srivastava, R. & Srivastava, B.S. (2002) Selective identification of new therapeutic targets of *Mycobacterium tuberculosis* by IVIAT approach. *Tuberculosis*, 82, 175-182.

[8] Hang, L.M., John, M., Asaduzzaman, M., Bridges, F. A., Vanderspurt, C., Kim, T.J., Taylor, R.K., Hillman, J.D., Progulske-Fox, A., Handfield, M., Ryan, E.T., & Calderwood, S.B. (2003) Use of *in vivo* induced antigen technology (IVIAT) to identify genes uniquely expressed during human infection with *Vibrio cholerae*. *Proc. Natl. Acad. Sci. USA* 100, 8508-8513.

[9] Kim, Y.R., Lee, S.E., Kim, C.M., Kim, S.Y., Shin, E.K., Shin, D.H., Chung, S.S., Choy, H.E., Progulske-Fox, A., Hillman, J.D., Handfield, M. & Rhee, J.H. (2003) Characterization and pathogenic influence of *Vibrio vulnificus* antigens preferentially expressed in septicemic patients. *Infect. Immun.* 71, 5461-5471.

[10] Cheng, S., Claney, C.J., Checkley, M.A., Handfield, M., Hillman, J.D., Progulske-Fox, A., Lewin, A.S., Fidel, P.L. & Nguyen, M.H. (2003) Identification of *Candida albicans* genes induced during thrush offers insight into pathogenesis . *Mol. Microbiol.* 48, 1275-1288.

[11] Rollins, S.M., Peppercorn, A., Hang, L., Hillman, J.D., Calderwood, S.B., Handfield, M. & Ryan, E.T. (2005) *In vivo* induced antigen technology (IVIAT). *Cell. Microbiol.* 7, 1-9.

[12] Hyush, T.V, Young, R.A. & Davis, R.W. Constructing and screening cDNA libraries in λgt10 and λgt11. In: Glover, D.M., editor, *DNA Cloning*, England, IRL Press, 1985, 49-78.

[13] Harlow, E. & Lane, D. Antibodies: *a Laboratory Manual.* Cold Spring Harbor, New York, Cold Spring Harbor Laboratory, 1988.

[14] Akhtar, P., Sriastava, S., Srivastava, A., Srivastava, M., Srivastava, B.S. & Srivastava, R. (2006) Rv3303c of *Mycobacterium tuberculosis* protects tubercle bacilli against oxidative stress *in vivo* and contributes to virulence in mice. *Microbes and Infection.* 8, 2855-2862.

[15] Agalicho, M.J. & Camilli, A. (2002) *In vivo* expression technology. *Infect. Immun.* 70, 6518-6523.

[16] Srivastava, V., Carine, R., Srivastava, R., Ramalingam, B., Locht, C. & Srivastava, B.S. (2007) Macrophages-specific *Mycobacterium tuberculosis* genes: identification by green fluorescent protein and kanamycin resistance selection. *Microbiology*, 153, 659-666.

[17] Srivasava, V., Jain, A., Srivastava, B.S. & Srivastava, R. (2008) Selection of genes of *Mycobacterium tuberculosis* upregulated during residence in lungs of infected mice. *Tuberculosis*, 88, 171-177.

[18] Saxena, A., Srivastava, V., Srivastava, R. & Srivastava, B.S. (2008). Identification of genes of *Mycobacterium tuberculosis* upregulated during anaerobic persistence by fluorescence and kanamycin resistance selection. *Tuberculosis*, 88, 518-525.

In: Recent Trends in Biotechnology, Volume 1
Editors: M.P. Singh, A. Agrawal and B. Sharma
ISBN 978-1-60876-148-7

Chapter 11

IMMOBILIZATION TECHNOLOGY: THE ACTUAL TREND IN BIOTECHNOLOGY

R. Sharma[1] and R. Singh[2]
[1] Department of Biotechnology, VBS Purvanchal University, Jaunpur-222 001, India
[2] Department of Biological Sciences, M.G. Chitrakoot Gramodaya Vishwavidyalaya, Satna- 485 331, India

ABSTRACT

The immobilization of enzyme and cells is a practical method when bioprocess employing biocatalysts are considered. At present, the application of immobilized biocatalyst and cells in bioprocessing include production of energy, drugs, and enzymes, valuable proteins, specialty products etc. pollutant treatment, biosensors with high sensitivity and specificity through various techniques of immobilization protocol, numerous numbers of matrix materials were tried. Recent advancement in sophisticated instrumentation and recombinant DNA technology added new dimension to this technology not only considering the biological, biochemical aspects, but also of economical factors engineering aspects such as mass transfer resistance and diffusional limitations when dealing with this technology at commercial scale.

INTRODUCTION

In a biochemical context, immobilization is a generic term to describe the retention of biologically active catalyst/cell organelles or even whole organisms (it may be wild or recombinant bacterial. fungal, plant and animals) cells to achieve the special bioprocess objectives. This technology involves multidisciplinary approaches like chemistry, biochemistry and cell biology on one hand with biochemical and process engineering on the other.

Single step reaction will be achieved by simply immobilizing the specific enzymes which are easier to prepare and are always the method of choice for simple, single enzyme systems

[1, 2]. However, in the case of multistage reaction immobilized whole cells are the only practical choice [3, 4].

All enzymes originate in whole cells and are in principle available at all levels of purity, from non-metabolizing cells through crystalline protein. The starting point for most enzyme immobilization at commercial level is normally a soluble protein purchased as a powder from a laboratory supplier that not only enhances the cost of the process but availability of large quantities of enzymes of a particular purity is doubtful. The academician while handling the whole sequence of processes from cell growth to immobilization and so has greater liberty to take purification to any extent and it is reported that purification not only affect the activity but also stability [5] and these two not always increase with additional processing. Intracellular enzymes in general seem to be stabilized by the proximity of cellular structure or inclusion in cell membranes, even when such structures or membranes are no longer alive or functional. It is well known that immobilization increases the stability of enzymes [6]. One of the reasons is a more stable structure due to physical and chemical interactions with the carrier structure [7]. The intraparticle diffusion resistance sometimes enhances the stability of the enzymes [8].

Immobilized whole cells are the only practical choice while for performing multistage reaction [3]. Immobilized microorganisms can be classified into two categories- dead cells and living cells. The latter is subdivided again into resting cells and dividing cells [9]. Immobilized dead cells use enzymes exuded from cell debris and retained inside carrier. They behave similarly to immobilize enzyme although cellular membranes may cause complications in transport phenomena [10]. Immobilized resting cells are characterized by homogeneous distribution inside the particles. This will happen where cells do not grow after immobilization. If the cells grow, the cell density distribution is not homogeneous. Usually the cell density becomes larger near the surface due to least mass transfer resistances. The cell growth must be regulated in this case so that the grown cells do not leak into the medium [11].

Intraparticle diffusion effects on the reactivity of immobilized microorganisms may be investigated similar to the case of immobilized enzymes systems. The principal difference is the existence of the cells in the biocatalyst domain. When the substrate reaches the surface of the cells, bioconversion is initiated. Thus, the substrate has to migrate through the space between the cells in order to react. This situation resembles diffusion through solid catalyst particles. Klein *et al.* presented the concept that the effective diffusivity, D_e, depends on the cell concentration in the gel, i.e. the diffusivity decreased with increasing cell concentration [12, 13]. Furusaki *et al.* [14] formulated this property using random pore theory of Wako and Smith [15]. Thus, D_e, is related to the diffusivity in the gel, D_g:

$$D_e/D_g = \varepsilon_c^2 \tag{1}$$

Where, ε_c is the volume fraction of gel occupied by cells. If the specific volume of cells, v_c, and the cell concentration c_c is used, the following equation holds using a consistent unit system.

$$D_e/D_g = (1-v_c c_c)^2 \tag{2}$$

The value of D_g is independent of gel concentration within the range of the experimental condition. It is in general 60-80% of the diffusivity in water equation 2 is used for the analysis of resting cells as well as growing cells.

As far as support matrix or carrier is concerned it is the most important decision to be made while performing the immobilization task. Every conceivable form of material could be applied to enzyme immobilization, but in general the choice is between particles, sheets, fibers [16] or membrane systems [11]. The bulk flow of substrate through a reactor occurs principally in the interstices between the biocatalyst elements and it is the dimension of these gaps which dictate overall fluid movement. However, transfer of substrate from the bulk fluid to active site relies on diffusion which significantly reduces the flux of substrate reaching active site within a porous material leading to diffusional limitation of mass transfer [17]. For academic studies, soft gels requiring complex chemical reactions and the limitation of flow are not a serious problem. While in industrial processing good stability and resistance to compression can override all other considerations.

The ideal matrix should have the following properties:

(i) Low cost
(ii) Chemically inert i.e. it should not react with substrate, product or any other components of the systems
(iii) Physical strength.
(iv) Stability
(v) Regenerability
(vi) Enhancement of enzyme speficity
(vii) Reduction in product inhibition.

Finally, unless there is a very specific requirement, it is wise decision to procure the matrix available commercially of high quality. The wide range of support matrices are available generally at ready to use level depending on technique of immobilization used. The immobilization techniques varies from very simple adsorption [18] to complex polymeric gel systems [19], which is one of the most popular and widely used techniques either for entrapping/encapsulating living whole cells or enzymes [20, 21] or co-immobilized systems (enzyme/enzyme, cell/cell, enzyme/cell) [18, 21] involving polymerization of monomers into polymeric gel system, and monomeric units can be purchased and one can prepare the gel with variable choice and flexibility.

TECHNIQUES OF IMMOBILIZATION

All the immobilization techniques typically reduce the activity by 10 to 50 percent, but this is not a big problem even if half of the activity lost can be overcome by using enzyme twice, to get it back and if it is used 100 times, the loss is meaningless. It is impossible to make a generalization about what type of enzymes are best suited for which techniques? Most of the enzymes are functional at physiological state hence, initially try for gentle technique than go for harsher technique [22]. The major problems and great concern associated with immobilized enzymes is microbial contamination and extra care should be taken to minimize the chance of contamination. The most serious problem involves membrane-bound enzymes, multienzyme systems and enzymes that require cofactors. Such system is particularly difficult to work with, and have, in most cases required the use of immobilized cells. The cells, whether living or dead, can keep the enzymes in the proper spatial arrangement with respect to each other and can regenerate cofactors [23].

Enzyme and cells can be immobilized on substances ranging from paper, wood chips and crushed red bricks through conventional ion exchange resins to sophisticated and costly ceramic and glass beads. The most widely used immobilization technique can be roughly divided according to whether the biocatalyst is chemically bound or physically retained but the classification is not clear cut. For example, adsorptive binding (caused by Van der Waals' forces) is grouped under carrier-bound preparations. It must be further emphasized, that combinations of the different classes of techniques occur. The following classification shown below in Figure 1 can be implemented to both the systems i.e. for enzymes as well as cells [23].

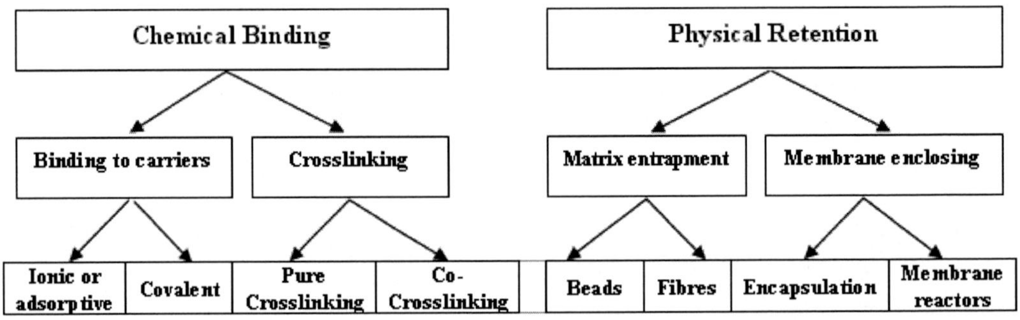

Figure 1. Major classes of Immobilized biocatalyst/cells.

A brief overview of the above mentioned techniques are described to reveal the basic understanding of the immobilization techniques.

ADSORPTION ON SOLID SUPPORT

Adsorption is used to attach the biocatalyst/cells to a surface by non-covalent interactions. The type of adsorption can range from non-specific binding dependent on weak surface forces, through the strong interactions of ionic bonds to the highly specific, multipoint binding of natural biological affinities [24]. This technique involves strength and sophistication of forces, although for any particular enzyme the presence or absence of the required characteristic will dictate the strength of the specific interaction. However, in practice a large number of enzymes have been immobilized by adsorption. All the techniques of adsorption involves with the outer shell of the enzymes (surface amino acids or sugar residue or cell wall components of the cells). Simple adsorption involves hydrogen bonding and Van der Waals' forces alone, is seldom found sufficient to hold enzymes and cells to a surface during catalytic operations but the combined effect of many weak interactions can prove very effective at holding polymeric materials on a surface. The reduction in surface free energy which results from adhesion of the cell wall to the solid is probably the major driving force of the initial adsorption, adhesion being optimum when the energy of the surface matches that of cell wall. Nutrient may also be concentrated at the surface by the same phenomenon, explaining the advantage to cells of adhering at low nutrient levels. If the surface is also biocompatible then rapid colonial growth may follow the introduction of the solid into a growing population of cells. Truly hydrophobic surfaces such as oil based plastics, rarely provides adequate capacity for practical immobilization of enzymes and cells, although some specialized applications do occur in immunological assays (e.g. ELISA) where very small quantities of protein antigen are adsorbed on the walls of microtitre plate [25]. Hydrophilic polymers to which hydrophobic ligands are bound are very effective for immobilizing enzymes. It is normally found that cells tend to interact best with anionic resins reflecting the acidic character of the outermost cell walls but morphological changes may follow the initial adsorption increasing the extent of binding. Certain phenolic resins like tannins and polyphenols are often found associated with enzymatic proteins in plants and soil. This affinity exploited in preparing a range of very versatile immobilized enzymes [26].

Covalent Attachment to Support

This technique is more useful for enzymes than for cells. Aqueous media and several factors which require some organic solvents are used to link a biocatalyst to support. In most circumstances adsorption or entrapment is likely to provide a more efficient and expensive method of immobilizing proteins. Most of the literatures [27] available in this technique are devoted to activation of the matrix hence it is important to decide first that which support to use rather than which activation method to apply. Many natural supports ranging from natural to synthetic, organic to inorganic are available.

Support matrix and their activation reaction involves very huge list, some of the noteworthy names are given below from each category just to familiarize with activation reaction and supporting matrix as well.

- Hydroxy Polymers- Carbonates and Imidocarbonates, Active esters, Epoxides, Benzoquinones, Triazines [28, 29] etc.
- Aldehyde Matrices- Oxidation of alcohols, Bis-aldehyde.
- Polyacids Matrices- Carbidiimide, Acid chloride, Azide.
- Polyamine Matrices.
- Thiol Matrices.
- Natural Organic Matrices- Polysaccharides (Agarose, Cellulose, Dextran & starches, Chitin/Chitosan), Proteins (Collagen, other soluble & insoluble proteins) [29, 30, 31].
- Inorganic Matrices- Organosilane, Porous/low porosity inorganics.
- Synthetic & condensed Polymers.

Entrapment in Polymeric Gels

Most biocatalysts are significantly larger in size than the substrate molecule on which they act. Therefore, it is possible to selectively retain an enzyme or cell behind a porous barrier while important solutes diffuse with relative freedom between the bulk fluid and the fixed enzymes entrapment does not result in a direct link between and active components rather confined to a small volume of liquid within a defined environment [10, 11]. Although the reaction which generates the porous barrier may coincidentally also cause some loss of activity. The main problem with entrapped enzyme and cells generally posed diffusional restrictions [17]. To be an effective barrier a matrix must have pores within a limited size range, being sufficiently small to retain the catalyst while not hindering the free diffusion of substrate and product. Entrapment technique broadly classified into two main categories: those in which a single membrane separate the enzymes solution from the bulk, and those in which three dimensional gel entrap the enzymes. The membranes held within a special reactor. The creation of a covalent bond between the biocatalyst and a support matrix forms a stable conjugate. The outer shell of all enzymes and cells presents an array of reactive functional groups. Almost every chemical reaction which can be used in module or the formation of an interfacial membrane around an emulsified mixture. Three dimensional gels are almost exclusively formed *in situ* with the biocatalyst distributed homogeneously throughout the matrix. Immobilization through entrapment has provided the greatest scope for deploying biocatalyst in different physical forms. Although beads packed in column reactor are common [32, 33], fibers, gel, sheets and emulsion can be used in fludized bed [34]. Gel beads often suffer from poor rigidity due to their very porous nature, while membrane rectors involve high cost with very less flexibility [35]. In practice membrane reactors are used for enzymes acting against low molecular weight substances, while three dimensional gels are widely used for immobilizing particulate enzymes or living cells. The main division in this group of technique using hydrogels is between those forms a gel as result of polymerization reaction in which new covalent bonds are formed and those methods which create a gel from existing pre-polymers [20, 21].

Crosslinking

The enzyme activity of immobilized preparation prepared by crosslinking with bifunctional/multifunctional reagents will depend upon the degree of crosslinking [36]. If high concentrations of crosslinking agents are employed, more bonds will be formed which generally results in lower activities. The additional bonds formed at high concentrations not only block the entrance of the substrate to the active sites but also may crosslink the active site itself. While cells loose their viability and ability to catalyze multienzyme reactions limiting it to single enzyme reaction because of crosslinking reagents reacts not only with the cell surface, but also to an appreciable extent with the proteins in the interior of the cells.

MAIN ADVANTAGES AND DISADVANTAGES OF IMMOBILIZATION TECHNOLOGY

Advantages

i. Retention of the enzymes/cells in the reactor.
ii. High concentration of enzymes/cells in the reactor.
iii. Control of microenvironment.
iv. Quantitative and rapid removal of the enzymes/cells.
v. Easy separation of enzymes/cells from the solution.
vi. Repeated and reuse of the enzymes/cells.

Disadvantages

i. Loss of activity/viability.
ii. Mass transfer problems.
iii. Physical discrimination between enzymes/cells and fluid
iv. Prolonged operation.
v. Empirical nature of immobilization technology.

APPLICATIONS OF IMMOBILIZATION TECHNOLOGY

Immobilization technology has potential applications almost in each area ranging from chemical to pharmaceutical, health care to environmental clean up and clinical to fermentation technology and above all biosensor developments and it has marked effects on these areas. Some noteworthy selected industrial applications are given below.

Industrial Applications

Biosensors offer a broad range of applications for immobilized bacteria, other cells or enzymes [37]. In general, all types of cells ranging from bacteria to mammalian cells are used if they prove to have specific activities relating to a certain substance which is to be detected. Biosensors using immobilized cells available commercially are a few in comparison to enzymes and it needs more R & D for its potential applications. The creatinine sensor is one example of a hybrid sensor in clinical analysis for the online monitoring of creatinine during blood dialysis [38]. Another area using immobilization techniques is process with mammalian cells with monoclonal antibodies being one possible product using membrane or polymeric gel systems [39, 40, 41, 42]. A large scale process using immobilized cells of *rhodococcus* which proved to be an outstanding economical success is the biotechnological production of acrylamide, high fructose corn syrup, 6-amino pencillinic acid, L-aspartate, etc. and production of ethanol, vinegar, alkaline protease, phosphatase, lipase are already established on an industrial scale. As far as environment is concerned, in waste water treatment in trickling filter, activated sludge process, rotating disc contactors etc. are widely used.

CONCLUSION

The increasing complexity of the field of immobilization technology means that the success of forthcoming research and development will depend greatly on a multidisciplinary approach. It requires the early consideration, not only microbiological and biochemical aspects, but also of economical factors engineering aspects such as mass transfer resistance and diffusional limitations.

Recent developments in recombinant DNA technology greatly affected this area leading the design of organisms with improved enzyme profiles and specifically tailored individual enzymes. Such strains are used for both as a source of single enzyme and as complex (whole cells) biocatalysts. It could also be assumed that progression in down stream processing will make more enzymes available at a reasonable cost.

Another future aspect is the use of biomimetic catalysts e.g. artificial enzymes. They are regarded as being stable and highly selective. The properties of immobilized biomimetic catalysts will become one of the main topics in the field of bioprocess engineering in the near future.

REFERENCES

[1] Chibata, I. (1978). Immobilized enzymes-Research and development. John Wiley & Sons, Inc., New York. 237-264.

[2] Chibata, I.; Tosa, T. & Sato, T. (1983). Immobilized cells in the preparation of fine chemicals. *Adv. Biotechnol. Proced.* 10: 203-222.

[3] Chibata, I. & Wingard, L.B. (1983). Immobilized microbial cells. *Appl. Biotechnol. Bioengg.* 4: 53-151.

[4] Bisping, B. & Rehm, H.J. (1988). Multistep reactions with immobilized microorganisms. *Biotechnol. Appl. Biochem.* 10:87-98.

[5] Messing, R.A. (1975). Immobolized enzymes for industrial reactors. Academic Press, New York. 181-206.

[6] Kennedy, J.F.; Melo, E.H.M. & Jumel, K. (1990). Immobilized enzymes and cells. *Chem. Engg. Prog,* 7:81-89.

[7] Brummer, W.; Hennrich, N.; Klockow, M.; Lang, H. & Orth, H.D. (1972). Preparation and properties of carrier bound enzymes. *Eur. J. Biochem.* 25:129-135.

[8] Messing, R.A. (1970). Relationship of pore size and surface area to quantity of stabilized enzymes bound to glass. *Enzymol.* 39:12-40.

[9] Tanaka, A. & Nakajima, H. (1990). Application of immobilized growing cells. *Adv. Biochem. Engg. Biotechnol.* 42: 97-131.

[10] Chang, T.M.S. (1964). Semipermeables microcapsules. *Scienec.*146: 524-527.

[11] Chang, T.M.S. (1977). Encapsulation of enzymes, cell contents, cells, vaccines, antigens, anriserum, cofactors, hormones and proteins. In *Biomedical applications of immobilized enzymes and proteins*. ed. Chang,T.M.S. Plenum, New York. 1:69-89

[12] Klein, J. & Schara, P. (1981). Some engineering aspects of immobilized cells. *Appl. Biochem. Biotechnol.* 6: 91-99.

[13] Klein, J. & Kressdorn, B. (1986). Engineering aspects of encapsulated cells in polymeric gels. *Biotechnol. Lett.* 8: 739-744.

[14] Furusaki, S. & Seki, M. (1985). Engineering aspects of immobilized biocatalyst. *J. Chem. Engg. Japan*. 18:389-401.

[15] Wako, N. & Smith, J.M. (1962). Engineering analysis of immobilized systems.*Chem. Engg. Sci.* 17: 125-145.

[16] Quiocho, F.A. & Richards, F.M. (1964). Intermolecular crosslinking of a protein in the crystalline state. *Proc. Nat. Acad. Sci.* USA. 52: 833-839.

[17] Toda, K. & Sato, K. (1985). Studies on Diffusional limitations and mass transfer during immobilization of catalase. *J. Ferment. Technol.* 63: 251-257.

[18] Messing, R.A. (1874). Simultaneously immobilized glucose oxidase and catalase in controlled pore titania. *Biotechnol. Bioengg.* 16: 1419-1424.

[19] Bernfeld, P. & Wan, J. (1963). Antigen and enzyme made insoluble by entrapping them into lattices of synthetic polymers. *Science*. 142: 678-682.

[20] Danielson, B. & Mosbach, K. (1988). Immobilized enzyme and cells. *Methods Enzymology*, 137: 12-20.

[21] Rehm, H.J. & Omar, S.H. (1993). Special morphological and metabolic behavior of immobilized microorganisms. In: Biotechnol., Vol.1: Biological fundamentals (ed sahm, H.), VCH Verlag, Weinheim, New York. 255-260.

[22] Fukui, S.; Sonomoto, K. & Tanaka, A. (1987). Entrapment of biocatalyst with photocrosslinkable resin polymers and urethane resin plymers.*Methods Enzymol.* 135:230-252.

[23] Hartmeier, W. (1985). Immobilized biocatalysts- from simple to complex systems. *Trends in Biotechnol.* 3: 149-153.

[24] Sulkowski, E. & Lasskowski, M. (1974). Venom exonuclease (phosphodiesterase) immobilized on conconvalin-A-sepharose. *Biochem. Biophys. Res. Comm.*. 57:463-468.

[25] Campbell, D.H.; Luescher, E. & Lerman, L.S. (1951). Immunologic absorbents.I. isolation of antibody by means of a cellulose-protein antigen. *Proc. Nat. Acad. Sci.* USA. 37:575-578.

[26] Watanabe, T.; Mori, T.; Tosa, T. & Chibata, I. (1979). Immobilization of aminoacylase by adsorption to tannin immobilized on aminohexyl cellulose. *Biotechnol. Bioengg.* 21:477-486.

[27] Axen, R.; Porath, J. & Ernback, S. (1967). Chemical coupling of peptide and proteins to polysaccharides by means of cyanogen halides. *Nature*. 214:1302-1304.

[28] Kay, G. & Crook, E.M. (1967). Coupling of enzyme to cellulose using chloro-s-triazines. *Nature*. 216: 514-515.

[29] Veith, W. & Venkatasubramanian, R. (1976). Collagen immobilized systems. *Methods Enzymology*. Academic press, New York. 44: 243-263.

[30] Tosa, T.; Mori, T. & Chibata, I. (1969). Studies on continuous enzyme reactions. IV. Enzymatic properties of the DEAE-sephadex-aminoacylase complex. *Agric. Biol. Chem.* 33:1053-1059.

[31] Fukui, S & Tanaka, A. (1984). Applications biocatalysts immobilized by prepolymer methods. *Adv. Biochem. Engg. Biotechnol.* 29: 1-33.
[32] Veliky, I.A. & Jones, A. (1981). Bioconversion of digitoxigenin by immobilized plant cells in a column bioreactor. *Biotechnol. Lett.* 3:551-554.
[33] Ramirez, O.T. & Mutharasan, R. (1989). Physical immobilization characteristics of a hybridoma in a glass bead packed bed reactor. *Biotechnol. Bioengg.* 33: 1072-1076.
[34] Schugerl, K. (1989). Biofluidization: Application of the fluidization technique in biotechnology. *Can. J. Chem. Engg.* 67: 178-184.
[35] Stephanopoulos, G. (1993). *Biotechnology.* VCH Verlag, Weinheim, New York. 3: 476-714.
[36] Schetjer, A. & Bar-Eli, A. (1970). Preparation and properties of crosslinked water insoluble catalase. *Arch. Biochem. Biophys.* 136: 325-330.
[37] Danielsson, B. & Mosbach, K. (1988). *Methods enzymology.* Academic Press. Orlando, San Diego. 137: 3-14.
[38] Karube, I. (1988). Hybrid biosensors for clinical analysis and fermentation control. *Methods Enzymology.* Academic Press. Orlando, San Diego. 137: 131-138.
[39] Shirai, Y. & Sasaki, R. (1988). Continuous production of erythropoietin with immobilized animal cells. *Appl. Microbiol. Biotechnol.* 29: 544-549.
[40] Nilson, K. (1987). Methods for immobilizing animal cells. *Trends Biotechnol.* 5: 73-78.
[41] Bugarski, B. (1990). Design and operation of an air lift reactor for production of monoclonal antibodies by immobilized hybridoma cells. *Period. Biol.* 92:121-122.
[42] Iijima, S. & Mano, T. (1988). Immobilization of hybridoma cells with alginate and urethane polymer and improved monoclonal antibodies production. *Appl. Microbiol. Biotechnol.* 28: 572-576.

In: Recent Trends in Biotechnology, Volume 1
Editors: M.P. Singh, A. Agrawal and B. Sharma
ISBN 978-1-60876-148-7

Chapter 12

A BIOCATALYTIC APPROACH FOR PRODUCTION OF 7-AMINOCEPHALOSPORANIC ACID

V. K. Nigam[1*], K. Kundu[2] and P. Ghosh[1]
[1] Birla Institute of Scientific Research, Statue Circle,
Jaipur-302001, India
[2] Chemistry Section, MMV, Banaras Hindu University,
Varanasi-221005, India

ABSTRACT

Cephalosporin acylases are industrially important enzymes, which hydrolyze cephalosporins to 7-aminocephalosporanic acid (7-ACA), a key intermediate required for the production of most of the clinically used cephalosporin derivatives i.e. semisynthetic cephalosporins. Cephalosporin-C (CPC), an important β-lactam antibiotic in the international market is exclusively produced by aerobic fermentation as such exhibits negligible antimicrobial activity but substitutions on C_3 and C_7 positions of β-lactam ring along with other structures generate semisynthetic cephalosporins with diversified antimicrobial activity e.g. cefazolin, cefotoxime, cephamandole and cephaclor etc. The production of 7-ACA and CPC was extensively reviewed in the chapter with some of our findings using both free and immobilized microbial cells.

INTRODUCTION

7- Aminocephalosporanic acid (7-ACA) is an important intermediate for the production of semisynthetic cephalosporins which is usually obtained from the chemical deacylation of the natural β-lactam antibiotic cephalosporin-C (CPC). Chemical deacylation of CPC requires several expensive steps such as use of organic solvent, a super-low temperature reaction (-70^0C), and several other extreme parameters that raise certain environmental issues. In order to overcome the problems associated with chemical route, an enzyme mediated process has

* E-mail: nigam.bisr@gmail.com

been explored and several microorganisms have been reported for the production of 7-ACA. The enzymatic process is a more simplified, high yielding and environmental friendly technique as compared to the chemical route of 7-ACA synthesis.

The enzymatic production of 7-ACA from CPC takes place in two ways. In one way, which involves three steps: CPC is oxidatively deaminated by D-amino acid oxidase to ketoadipyl 7-ACA which later gives glutaryl 7-aminocephalosporanic acid in presence of hydrogen peroxide and finally glutaryl 7-aminocephalosporanic acid is converted to 7-ACA in presence of enzyme glutaryl 7-ACA acylase. The second pathway involves a single step conversion of cephalosporin-C to 7-ACA in presence of cephalosporin-C acylase enzyme.

Cephalosporin acylases are of two types [1-2]; 7-ß-(4-carboxybutanamido) cephalosporanic acid acylase, more popularly known as glutaryl 7-aminocephalosporanic acid acylase i.e. GL 7-ACA acylase, which hydrolyzes 7-ß-(4-carboxybutanamido) cephalosporanic acid, also known as GL 7-ACA to 7-ACA [3-10] and cephalosporin-C-acylase (CPC acylase) which hydrolyzes cephalosporin-C to 7-ACA [1, 11-14]. The enzymatic conversion of CPC to 7-ACA by different ways are shown in Figures 1 and 2 respectively.

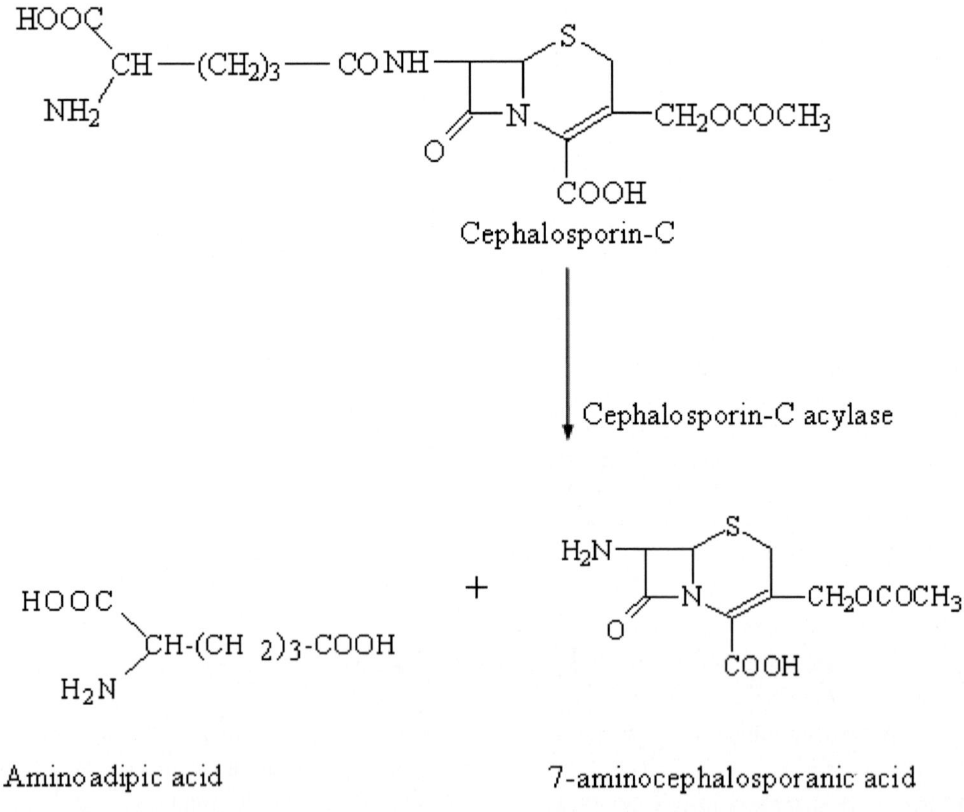

Figure 1. Single step deacylation of cephalosporin-C to 7-ACA.

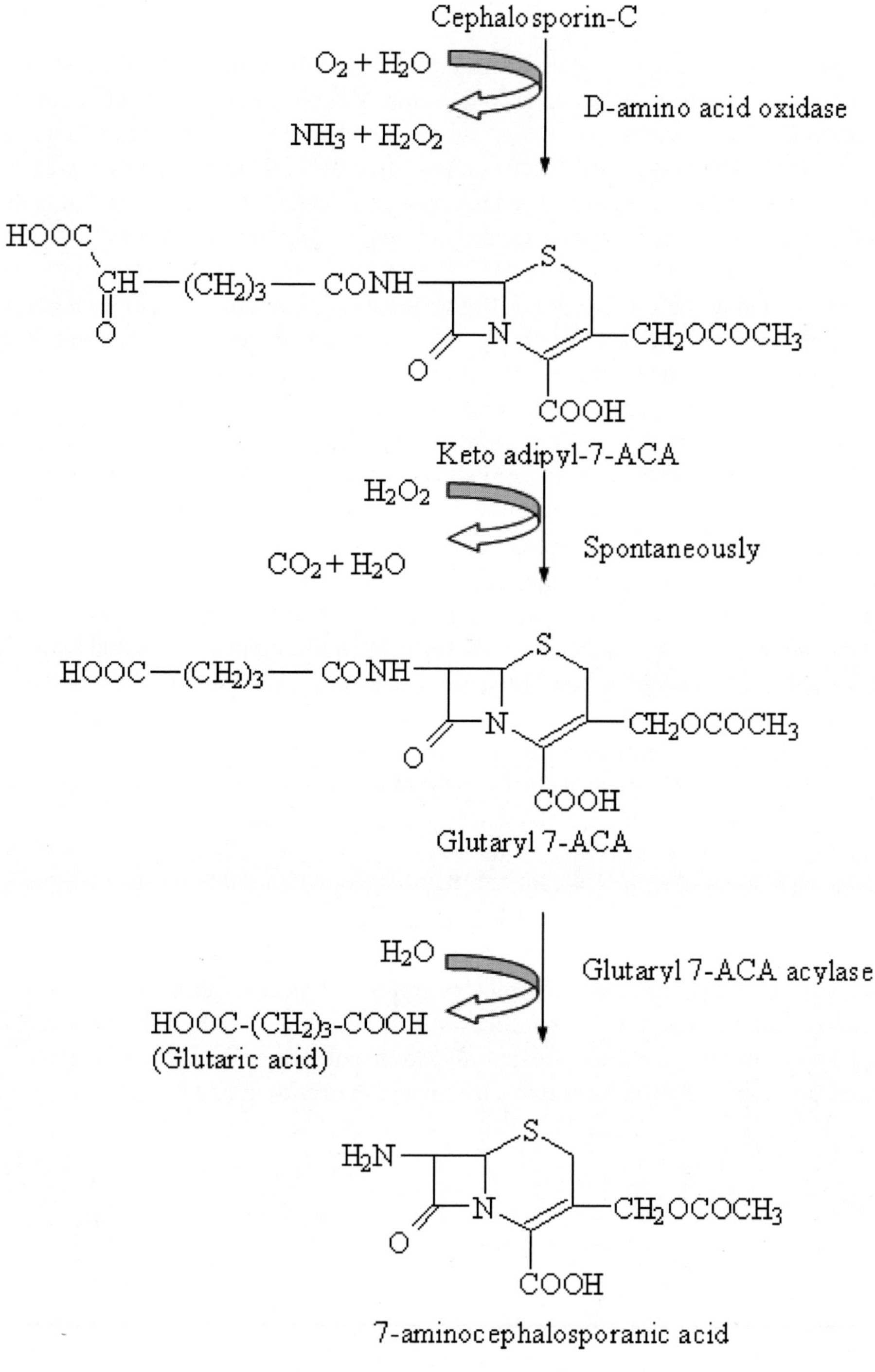

Figure 2. Three steps conversion of cephalosporin-C to 7-ACA.

HISTORY

The great modern advances in chemotherapy have come from the chance discovery that many microorganisms synthesize and excrete compounds which are selectively toxic to other microorganisms. These compounds, called antimicrobial compounds, have revolutionized the modern medicine. The first chemotherapeutically effective antibiotic was discovered in 1929 by Alexander Fleming, a British bacteriologist. He called that antibacterial substance 'Penicillin' because it is secreted by a species of mold, *Penicillium notatum*.

Antibiotics are chemical substances secreted by microorganisms which in low concentration inhibit the growth or kill the other microorganisms i.e. "antibiotics are antimicrobial agents of microbial origin" [15]. Microorganisms appear to be programmed to produce antibiotics only when the specific growth rate decreases below a certain level. This phenomenon probably was established during evolution in response to competitive pressures [16]. Antibiotics differ from each other in their chemical structures and the mechanisms of physiological action (with respect to composition of the side chain). On the basis of their range of action, the antibiotics may be classified into narrow range or broad range antibiotics. Some microorganisms grow to secrete antibiotics into the surrounding medium, others secrete only a part of antibiotics into the surrounding medium while the considerable amount of antibiotics remain bound with cells, and the third group of microorganisms do not secrete the antibiotics into the medium and the antibiotics remain almost completely bound with the cells of organism. Each species of microorganisms, due to a specific inherited property secrete a specific type of antibiotics. Antibiotics are the products of secondary metabolism which are produced during the stationary phase of the growth i.e. when most of the cellular growth has already occurred [17]. Secondary metabolites have also been termed "idiolites" because they are formed during the idiophase (production phase) of batch cultures [18].

Penicillin, having a benzyl side chain was the first generation of β-lactam antibiotics, which was later designated as penicillin-G. Rather than being a single substance, penicillin proved to be a class of compounds. The production of the particular groups of penicillin usually depends on the composition of the medium in which the mold was grown. Subsequently, in 1956 cephalosporins, another group of β-lactam antibiotics were discovered which are resistant to microbial penicillinases [19]. These have a four membered β-lactam ring fused to a six membered unsaturated dihydrothiazine ring. Penicillins have also a four membered β-lactam ring fused with a five membered thiazolidine ring (Figure 3).

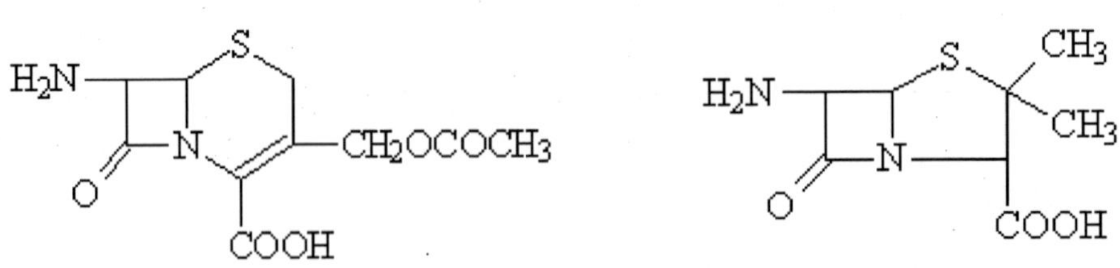

Figure 3. Structures of 7-ACA and 6-APA.

The β-lactam antibiotics are natural products and industrial production of all β-lactam antibiotics depends on microbial fermentation. The microbially synthesized penicillins and cephalosporins are called natural penicillins and natural cephalosporins respectively. These natural products can either be directly applied for clinical use or can be further modified through chemical or enzymatic methods to impart more valuable semisynthetic penicillins or semisynthetic cephalosporins.

Cephalosporium acremonium (*Acremonium chrysogenum*) is an industrially important fungus that produces ß-lactam antibiotic, cephalosporin-C (CPC) through fermentation [20-22]. Different strategies have been applied for the better production of CPC and also several methods have been developed to get higher yield of CPC [23-27]. The morphological differentiation of *Acremonium chrysogenum* during submerged fermentation of CPC was correlated with the productivity of antibiotic. It was observed that the production of CPC was highest when the number of swollen hyphae and the arthrospores were high. Also, the seed of early stage gave maximum yield [28]. The production of CPC was thoroughly investigated by varying the medium constituents and it was observed that a C/N ratio of 8.0, methionine at a concentration of 4 g/L and ammonium sulphate as inorganic nitrogen sources, respectively were most suitable for higher yield of CPC [29]. The fermentation of cephalosporin is a highly aerobic process. Oxygen deficiency results in increased penicillin-N synthesis. The fermentation is usually carried out at 28°C for 140 to 160 h in a batch mode. During the antibiotic fermentation, readily assembliable form of carbon (glucose) decreases antibiotic production. This is due to catabolite repression. Phosphate also inhibits the cephalosporin-C fermentation. It exerted its effect indirectly by regulating the rate of glucose consumption.

While the natural penicillins possess the advantages of high activity against many pathogenic bacterial strains and low toxicity towards the host, they suffer the disadvantages of being unstable in acid and sensitive to cleavage of the β-lactam by penicillinase, an enzyme produced by certain strains of bacteria. These defects in penicillins are the main cause for the discovery of semisynthetic penicillins possessing acid stability, penicillinase resistant and improved antibacterial activity. These are produced by the modification of natural penicillins.

Cephalosporin-C on the other hand, is relatively stable in acid solution and is resistant to penicillinase. Because of its weaker antibacterial activity, the modification of cephalosporin-C leads to the discovery of semisynthetic cephalosporins. The structural modification of these natural penicillins and cephalosporins mainly involve alternation of the substituents around the peripheries of the natural ring system [30].

7-aminocephalosporanic acid (7-ACA) is the key intermediate, essentially required for production of a number of clinically important cephem antibiotics. These semisynthetic cephalosporins are obtained by the cleavage of the side chain of cepholosporin-C [31]. The production of 7-aminocephalosporanic acid from cephalosporin-C takes place either by chemical methods i.e. using nitrosyl chloride [32] and imminoether [33] or by enzymatic methods.

The importance of these cephalosporin acylases in the pharmaceutical industry can be judged from the projected annual worldwide market of ~ 400 million US dollars. The greatest hindrance to enzymatic industrial production is that acylase takes GL-7-ACA as a primary substrate and its specificity (and activity) is too low for CPC [14].

The reports on fermentation of cephalosporin acylases are very few and are either in abstract form or are covered by patents. This is mainly due to its commercial importance. Alternate routes for production of 7-ACA have also been explored, but the reports are patented. D-amino acid transaminase isolated from *Bacillus licheniformis* ATCC 9945 and gamma glutamyl transpeptidase from *Pseudomonas* and *Bacillus* converts CPC to 7-ACA in presence of α-keto acids. In another approach CPC is converted to 7-phenoxyacetamido cephalosporanic acid (V-7-ACA) by acyl transferase of *Penicillium chrysogenum* and later cleavage of V-7-ACA by penicillin V acylase leads to the formation of 7-ACA [34].

The use of immobilized whole cells to study the microbial transformation is now well established both scientifically and industrially. It is generally impractical to reuse suspensions of native enzymes or other biocatalysts because their recovery from the reaction mixture is difficult and expensive. On the other hand, immobilization of whole cells allows both reuse of biocatalysts and their application in continuous process. The immobilized whole cells as biocatalysts have applications in food, pharmaceutical and chemical industries along with analytical and medical applications. The immobilized whole cells as biocatalysts have been used to produce several antibiotics such as bacitracin, ampicillin, cephalosporin-C, penicillin-G and 6- APA etc [35-37]. Very few reports are available on the biosynthesis of 7-ACA using immobilized systems.

CEPHALOSPORIN ACYLASES PRODUCING ORGANISMS

Cephalosporin acylases are industrially important enzymes which convert cephalosporins into 7-ACA, a key raw material for the preparation of semisynthetic cephalosporins. English *et al.* have reported acylase activity from *Aerobacter aerogenes, Proteus vulgaris, Pseudomonas aeruginosa* and *Pseudomonas phaseolicola* etc [38]. Batchelor *et al.* examined 160 fungi, 30 yeasts, and 25 actinomycetes using a mixture of natural penicillins as a substrate and reported 24, 3 and 11, respectively with acylase activity [39]. Of the 310 strains of *Escherichia coli* that Holt & Stewart tested, 125 were found with acylase activity [40]. Acylase activity is not restricted to microorganisms only as Cole observed that pig kidney acylase I hydrolyzed phenoxymethyl penicillin [41]. Cole & Sutherland cultured 148 strains of bacteria at 37°C and observed only one with acylase activity. Subsequently, the strains were cultured at 26°C in another medium containing phenyl acetic acid as one of the constituents, 10 were observed with acylase activity [42]. Acylase activity has been reported by Cole in fungi also belonging to the genera *Aspergillus, Cephalosporium, Epidermophyton, Penicillium* and *Trichophyton* [43].

Large scale surveys have indicated the rather small incidence of acylase producing microorganisms; nevertheless, this enzymatic activity is not restricted to a few genera, but appears to be distributed widely among the different genera of microorganisms. There are also indications that acylase is produced by individual strains and is not a property of the species. Until today no penicillin acylase has been reported that can remove the side chain of the natural cephalosporin-C in order to obtain 7-ACA. However, it is relevant that, although penicillin acylases surely are not cephalosporin-C acylases, many β-lactamases (penicillinases) display cephalosporinase activity [44-48].

Recently, various microorganisms have been reported for both cephalosporin-C acylase and glutaryl 7-ACA acylase activity [1, 8, 49-50]. Cephalosporin-C acylase is produced mostly by fungi and some bacteria. The organisms that have been reported to produce cephalosporin-C acylase are *Arthrobacter viscosus, Bacillus megaterium* ATCC 53667,

Paecilomyces C 2106, *Penicillium griseofulvin, Penicillium regulosum, Pseudomonas putida, Pseudomonas* sp. N 176 and V22 and *E. coli* [1, 50-54].

Reyes *et al.* reported extracellular cephalosporin-C acylase activity in fungi of the genera *Aspergillus, Fusarium* and *Penicillium* and examined that during the process of autolysis, its activity increased with incubation time and with increasing pH of the medium [51].

PRODUCTION OF CEPHALOSPORIN ACYLASES AND THEIR PHYSICO-CHEMICAL PROPERTIES

It has been reported that organisms require specific nutrients and unique environmental conditions for the production of enzymes, acylases. Kameda *et al.* proposed a selective isolation procedure based on growth on a mineral medium containing benzyl penicillin or phenyl acetic acid as sole source of carbon, and they correlated hydrolysis of acyl-DL-amino acid with the presence of penicillin acylase [55].

Walton used amide substrates in a soil enrichment procedure for isolating microorganisms that would liberate 7-aminocephalosporanic acid from cephalosporin-C [56]. Vigorous aeration and a pH range of 6.5-8.5 were required for good acylase activity when *E. coli* was grown in an inorganic salt medium added with either yeast extract and amino acids or peptides [57]. Acylase activity was found to be greatest when the cells were grown at 24°C, and very little activity was observed at 37°C. It has been examined that enzymes are synthesized at specific pH and temperature. For *E. coli* and other bacteria the most favorable temperature range for the synthesis of enzyme is 24°-26°C [42]. *P. rettgeri* does not grow at 37°C but both growth and enzyme synthesis are optimal at 20°C [58]. It was found that the addition of phenyl-acetic acid in growth medium increases the acylase production. Szentirmai reported that *E. coli* acylase production was increased when either phenoxy-acetic acid, 2, 4-dimethoxyphenylacetic acid, or p-methoxy-phenylacetic acid was added to the growth medium. The same author reported that readily metabolizable substances such as glucose, fructose and glycerol repressed acylase activity and that this repression decreased with decreased temperatures of incubation [59].

Similarly, the acylase activity of fungal cultures has been reported to be stimulated by the side chain precursors of the penicillins they hydrolyze. In this respect, Cole demonstrated that the addition of phenoxy-acetic acid to fungal cultures increased acylase titers [43]. Vandamme & Voets were able to isolate acylase producing microorganisms from soil using a mineral medium containing N-acetylglycine, N-glycylglycylglycine, or phenyl-acetamide as sole source of carbon and nitrogen [60]. Analysis of the results published in the form of papers and patents revealed that the following are mainly responsible for acylase production; (i) presence of complex organic source, (ii) presence of inducers, (iii) presence of repressors, and (iv) temperature, pH and aeration. Production of acylase takes place actively only when growth is sparse. For most of the bacteria, the nutrients are provided by complex organic sources such as yeast extract, malt extract, soyabean meal, corn steep liquor and hydrolyzed proteins or aminoacid mixtures [42, 61-63].

Shimizu *et al.* reported cephalosporin-C acylase activity of *Kluyvera citrophila* in medium containing both yeast extract and glutamate along with poly peptone [64]. Shibuya *et al.* reported acylase activity in strains of *Pseudomonas putida* ATCC 950 and *Pseudomonas*

SY-77-1 grown in a medium containing peptone, meat extract, yeast extract etc [3]. Ichikawa *et al.* reported different mutants of *Pseudomonas* strains, having acylase activity in two different production media. Medium A, containing milk casein etc. was used for strain NS-2 and G-2. Medium B, containing 3% extract Banito and 1% monosodium glutamate was used for the strains, Ci-36 and GK-16 [4]. The purification and properties of these mutant strains derived from *Pseudomonas* have also been described by Ichikawa *et al.* [5].

Table 1. Optimum pH and temperature for synthesis of cephalosporin acylase enzymes by different microorganisms

Organisms	Optimum pH	Optimum Temp. (°C)	References
Pseudomonas sp. A 14	Neutral	30	[1]
Bacillus laterosporus J1	--	30	
Pseudomonas sp. N176 & V22	Neutral	30	
Pseudomonas SY-77-1	7.5	30	[3]
Pseudomonas putida ATCC 950	7.5	30	
Kluyvera citrophila KY 7844	7.0	30	[64]
Kluyvera citrophila KY 3641	7.0	30	
Pseudomonas strains NS-2 &G-2	8.0	32	[4]
Pseudomonas strains Ci-36 & GK-16	8.0	25	[5]
Arthrobacter strain 45-8A	7.2	28	[31]
Bacillus sp.	7.2	28	[65]
Pseudomonas sp.	7.2	28	
Acremonium chysogenum Hm 178	7.0	30	[66]
Aspergillus sp.	5.0	25	[51]
Fusarium sp.	5.0	25	
Penicillium sp.	5.0	25	

Reyes *et al.* cultivated a number of fungi belonging to the different genera in the basal medium and observed extracellular CPC acylase activity [51]. Aramori *et al.* isolated two strains of glutrayl-7-ACA acylase producers and two strains of cephalosporin-C acylase producers after growing the different bacterial strains either on Bouillon agar plates or in meat extract broth (polypeptone 0.5%, sodium glutamate 0.5%, meat extract 0.2% and $MgSO_4 . 7H_2O$, 50 μg/mL) [1]. A cephalosporin acylase producing *Arthrobacter* strain 45-8A, hydrolyzing both cephalosporin-C and GL-7ACA to 7-ACA was isolated from soil by Zhang & Xu [31]. The strain was cultivated in peptone-yeast medium contained 14.0 g peptone, 5.0 g yeast extract, 3.0 g adipic acid, 10.0 g glucose and 10.0 g glutamic acid per liter. The production of 7-ACA in batch mode using different bacteria and characterization of their physicochemical properties were carried out by Nigam and Kundu [54]. The authors have compared the production of 7-ACA by three different CPC acylase producers and reported that *P. diminuta* offers certain advantages over *B. megaterium* and *E. coli* in terms of substrate specificity and thermal stability.

Table 1 shows the optimum pH and temperature for the synthesis of acylase enzyme by different microorganisms.

CLASSIFICATION OF CEPHALOSPORIN ACYLASES

Cephalosporin acylases are general name for the enzymes which deacylate cephalosporins to 7-aminocephalosporanic acid. Later, a group of cephalosporin acylases named glutaryl 7-ACA acylase were found which could deacylate glutaryl 7-ACA and other cephalosporins possessing an aliphatic dicarboxylic acid as the 7-position substituent [3, 5]. Glutaryl 7-ACA is important intermediate from the point of view of industrial production of 7-ACA; the starting material for the manufacturing semisynthetic cephalosporins. The classification of cephalosporin acylase is summarized in Figure 4 along with their substrate specificities.

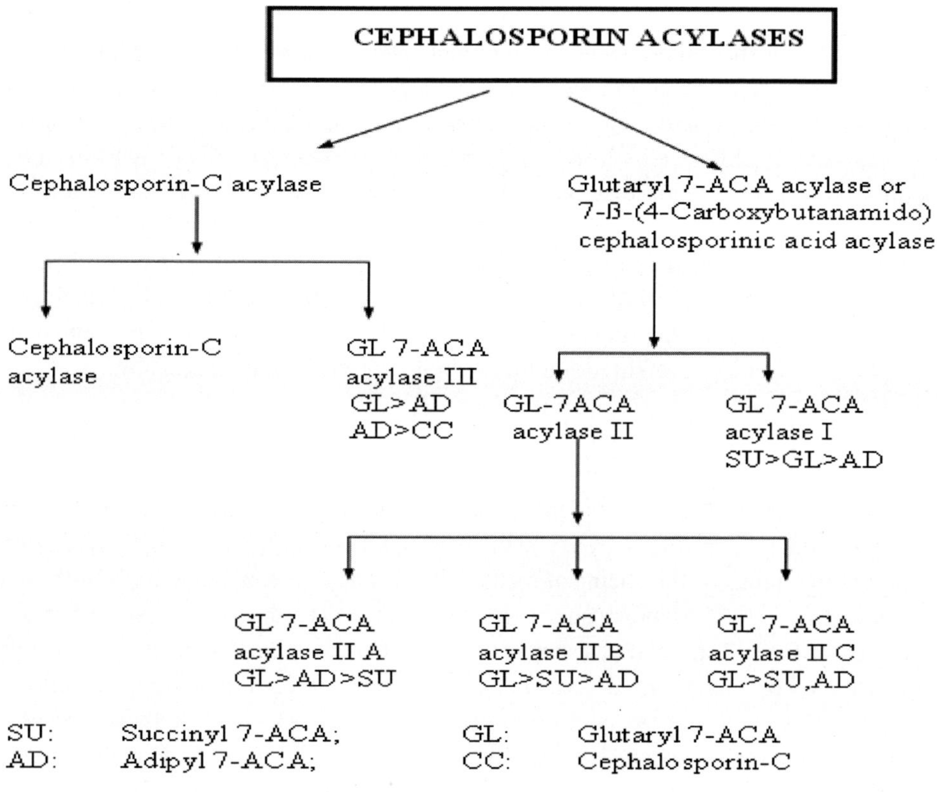

Figure 4. Classification of cephalosporin acylases.

SUBSTRATE SPECIFICITY OF CEPHALOSPORIN ACYLASES

Like penicillin acylases, the substrate specificity of cephalosporin acylases is determined by the side chain structure and not by β-lactam nucleus. Cephalosporin acylases hydrolyze succinyl 7-ACA, glutaryl 7-ACA, adipyl 7-ACA and cephalosporin-C to different extent. The degree of hydrolysis of side chain structures indicates that increase in the length of the side chain by one-CH_2- (i.e. glutaryl to adipyl) decreases the rate of hydrolysis by 30-75% whereas introduction of –CH_2-NH- (i.e. glutaryl to aminoadipyl) decreases the rate of hydrolysis by about 95%. Glutaryl 7-ACA acylases hydrolyze glutaryl, succinyl and adipyl 7-ACA but not cephalosporin-C. The rate of hydrolysis varies from one enzyme preparation to the other. Cephalosporin-C molecules contain side chain amide bond between delta-carboxylic acid group of D-amino acid and 7-amino group of the 7-ACA. Enzyme (gamma-glutamyl transpeptidase) hydrolyzing the gamma carboxylic acid amide of L-glutamate (structure similar to side chain of succinyl 7-ACA) is known and widespread. Further, Lowe observed that obstruction for enzymatic hydrolysis of cephalosporin-C is due to terminal D-amino acid structure in cephalosporin-C molecule [67].

PRODUCTION OF 7-AMINOCEPHALOSPORANIC ACID (7-ACA) FROM CEPHALOSPORIN-C

7-Aminocephalosporanic acid is an important intermediate, essential for the production of clinically used cephem anitibiotics, i.e. semisynthetic cephalosporins. The search for cephalosporin acylases was driven by the fact that enzymatic conversion utilizing penicillin acylase was successfully applied to the preparation of 6-aminopenicillanic acid from benzyl penicillin. Cephalosporin acylase activity was reported by various workers however, first true report on enzymatic deacylation of cephalosporin-C forming 7-ACA was given by Walton [56]. This compound is generally prepared by removing side chain of naturally occurring cephalosporin-C. Cephalosporin-C as such exhibits negligible antimicrobial activity, but substitutions at C_3 and C_7 positions with other structures generates semisynthetic cephalosporins with diversified antimicrobial activity, e.g. cefazolin, cefotaxime, cephamandole, cephaclor etc. [68].

Since, chemical methods of 7-ACA production using imminoether and nitrosyl chloride involve multiple cost consuming steps, highly reactive chemicals which can only be handled at low temperatures and require expensive safety measures and energy supply, as well as strong environmental controls, an enzymatic conversion method has been long explored to overcome shortcomings of the chemical methods [69]. The two enzymatic process/three steps for production of 7-ACA is more successful as compared to the single step conversion of CPC to 7-ACA. Table 2 summarizes some examples of semisynthetic cephalosporins [70].

Frederiksen & Emborg described another approach involving conversion of cephalosporin-C to 7-phenoxyacetamido cephalosporanic acid (V-7-ACA) by acyl transferase of *Penicillicum chrysogenum* and subsequent cleavage of V-7-ACA by penicillin-V acylase to 7-ACA [34]. A number of other approaches are there for biosynthesis of 7-ACA and as the product is of great pharmaceutical importance, the process of searching new isolates with high CPC acylase capacity, genetic manipulation in DNA molecules of producing strains are still in progress. 7-aminodeacetylcephalosporanic acid (7-ADACA) is another important starting molecule for the synthesis of clinically important oral cephem antibiotics such as cefixime. 7-ADACA is derived from cephalosporin-C via 7-ACA.

Table 2. Some important semisynthetic cephalosporins

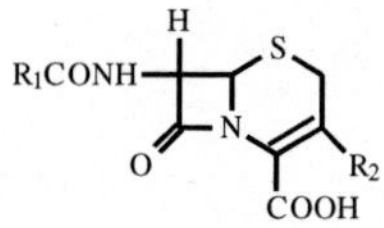

Name	R_1	R_2
1. Cephaloridine	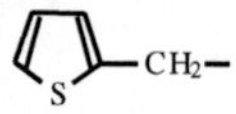	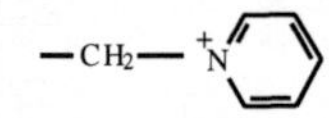
2. Cephalothine	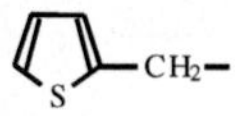	$-CH_2-OCOCH_3$
3. Cephalexine	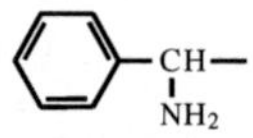	$-CH_3$
4. Cephamandole	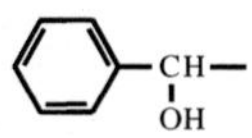	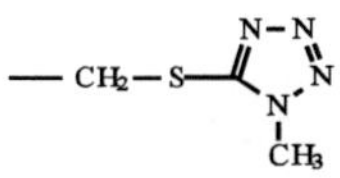
5. Celsulodin		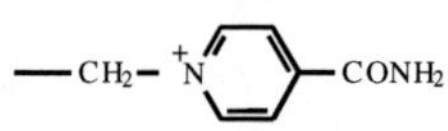
6. Cefoperazone	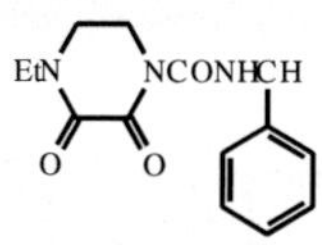	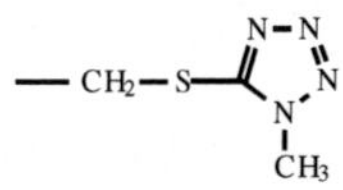
7. Cefotaxime	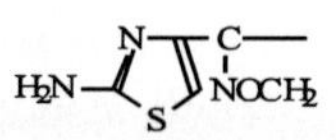	$-CH_2-OCOCH_3$
8. Ceftazidime	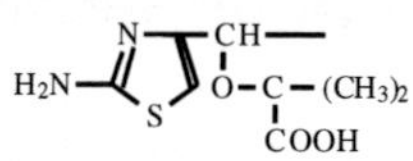	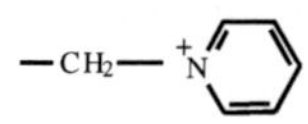
9. Cefuroxime	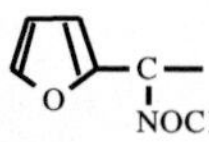	$-CH_2-OCONH_2$
10. Cefazoline	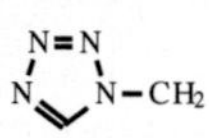	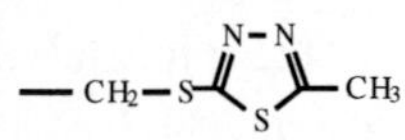
11. Ceforanide	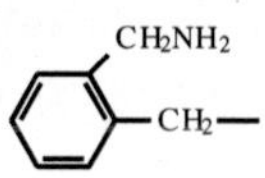	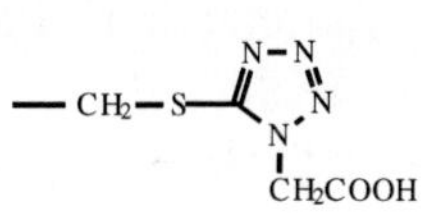
12. Cefotiam	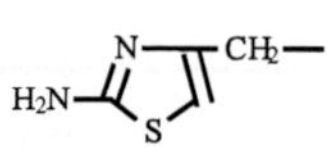	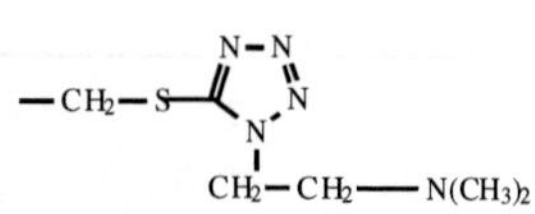
13. Cephaloglycin	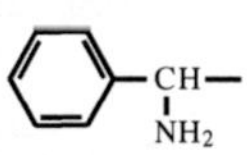	$-CH_2-OCOCH_3$

Table 2. (Continued)

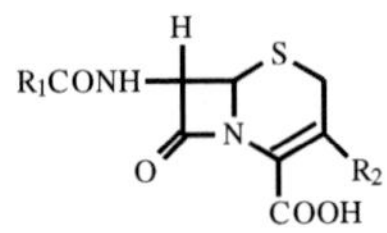

Name	R1	R2
14. Cefatriazine	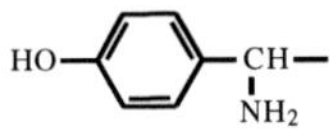	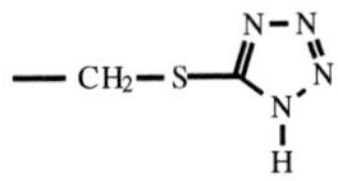

The direct production of 7-ACA by *Acremonium chrysogenum* Hm178 transformated with a 7-ACA biosynthetic operon, constructed from two enzyme genes, D-amino acid oxidase (DAO) from *Fusarium solani* MO718 and cephalosporin acylase from *P. diminuta* V22 have been established [71]. Single step conversion of cephalosporin-C to 7-ACA has been demonstrated by various workers [1, 11, 12, 31, 56], but the process of single step enzymatic conversion of CPC to 7-ACA is not popular because of non-availability of hyper-producing cultures of cephalosporin-C acylase. Most of the work on the production of 7-ACA is restricted to the two enzymatic pathways and a number of reports are available on the assessment of D-amino acid oxidase (DAO) that catalyses conversion of CPC to GL-7-ACA and GL-7-ACA acylase. D-amino acid oxidase (EC 1.4.3.3) catalyzes the oxidative deamination of a variety of D-amino acids with the production of respective α-keto acid and ammonia, an oxidative process that is applicable in the 7-ACA biosynthetic route [72-78].

Immobilized whole cells have been in use for the production of most of the commercially important compounds. Several antibiotics have also been produced from the immobilized microbial cells. The single step conversion of cephalosporin-C to 7-ACA by immobilized cephalosporin-C acylases have not been reported till date, yet the three steps of conversion by immobilized enzymes are there in literature.

Dey *et al.* reported that activity of D-amino acid oxidase from *T. variabilis*, involved in the conversion of cephalosporin-C to GL 7-ACA, by immobilizing on different supports (manganese oxide, magnetic iron oxide particles etc.) remained above 100% during the first 9 hours of a semi-continuous conversion of cephalosporin-C [79]. Tischer *et al.* have also investigated the biocatalytic production of 7-ACA by immobilizing both D-amino acid oxidase and glutaryl 7-ACA acylase. 7-ACA was obtained in a yield of 94-95% with a final substrate concentration of about 4% at the end of the reaction. The immobilized D-amino acid oxidase and Glutaryl 7-ACA acylase reached half activity after 138 and at least 150 cycles, respectively [11].

Conversion of CPC to GL 7-ACA by D-amino acid oxidase from *Trigonopsis variabilis*, covalently immobilized on the polysterenic resin duolite was carried out by Alfani *et al.* in both batch stirred and fluidized bed reactors [80]. Similarly, an economical process for the enzymatic oxidation of cephalosporin-C to glutaryl 7-ACA at a pilot plant scale utilizing whole cells of the yeast *Trigonopsis variabilis* containing high levels of D-amino acid oxidase has been successfully established [78].

Nigam *et al.* evaluated the production of 7-ACA by free and immobilized whole cells *of P. diminuta* in different carriers (chitosan, gelatin and agar respectively) and noticed that immobilized whole cells can be reused for several cycles without loosing much of the CPC acylase activity. The performance of entrapped cells showed that chitosan was the best immobilizing support for the production of 7-ACA. They also observed that the yield of 7-ACA was higher with lysozyme treated cells as compared to sonicated and whole (intact) cells [81-82]. The same author also investigated the continuous production of 7-ACA using immobilized cells in a packed bed tubular column and reported that cell to carrier ratio of 0.45 and a flow rate of 120 mL/h were most suitable for the conversion of CPC to 7-ACA. Operational stability was highest with chitosan (half life 2106 h) followed by gelatin and agar as immobilizing supports [83].

Although, many attempts have been made to substitute the chemical process of 7-ACA production by a biological process, yet not a single satisfactory bio-catalytic process is commercialized for 7-ACA production in large scale. It has been reported by various workers that acylases also promote acylation reaction i.e. the reactions catalyzed by acylases are reversible. The hydrolytic reaction is catalysed usually at alkaline pH (7.0-8.5) while acylation reactions occur at acidic or neutral pH (4.0-7.0) [84-86, 64].

EFFECT OF VARIOUS COMPOUNDS (INDUCERS AND REPRESSORS) ON PRODUCTION OF 7-ACA

It has been observed that addition of certain substances/compounds (side chain analogs) to the production media; induce the synthesis of enzyme CPC acylase and thus the production of 7-ACA increases. The effect of glutaric acid and adipic acid, the side chain analogs was observed to increase the CPC acylase activity and thus production of 7-ACA [3, 4 and 65]. It was also observed that the maximum extent of stimulation occurred when inducers were added at certain time of growth phase.

At the same time it was also found that acylase synthesis is catabolically repressed by readily metabolizable carbon source, glucose. Shimizu *et al.* reported that glucose and glycerol repressed the production of acylase. They studied the effect of glucose and c-AMP on acylase production by *K. citrophila* KY 7844 cells and found that glucose at even very low concentration repressed the acylase activity. The catabolite repression by glucose was released to some extent by additon of 3', 5'-cyclic AMP (c-AMP) into the culture media [64]. In another studies, it was reported that glucose at a concentration of 0.2% decrease the production of acylase enzyme to 22% in *E. coli* JM 109 [7]. The effect of different organic compounds on the biosynthesis of cephalosporin-C acylase involved in a single step conversion of CPC to 7-ACA showed that glutaric acid and adipic acid at a concentration of 0.2 g/L act as inducers while glucose and glycerol at a concentration above 1.5 g/L act as repressor for the enzyme synthesis [53]. The time course of addition of inducer in the production medium was also investigated and it was found that the best induction occurred when inducer was added at 9 h of the growth of microorganisms.

Molecular Structure of Cephalosporin Acylases

Studies relating to physicochemical characterization and molecular aspects of these enzymes had been far fewer and thus could be attributed to the commercial importance and application of these enzymes. The acylases from different sources have been cloned and complete amino acid sequences have been deduced from nucleotide sequence, coding for either the precursor or the active enzyme. Molecular aspects of these enzymes have been reviewed and summarized in Table 3 [2, 12].

Table 3. Molecular weights and subunits structure of cephalosporin acylases

Enzyme	**Source**	**Molecular weight**	**Subunits**		**References**
			Alpha	**Beta**	
Cephalosporin-C acylase					
Cephalosporin-C acylase I	*E.coli*[a] JM109	84000	26000	58000	[7]
Cephalosporin-C acylase	*E.coli*[b] JM 109	84000	26000	58000	
GL 7-ACA acylase II	*Pseudomonas* SE 83	83600	25400	58200	[49, 87]
GL 7-ACA acylase					
GL 7-ACA acylase	*Pseudomonas* *GK*16	70000	16000	54000	[6]
GL7-ACA acylase I	*Pseudomonas* SE 83	58100	38200	19900	[87]
GL 7-ACA acylase	*E.coli*[c] MC 1061	89000	28000	61000	[7]
GL 7-ACA acylase	*E.coli*[d] JM 109	70000	--	--	[8]
GL 7-ACA acylase I	*B. subtilis*[d]	70000	--	--	

Parent organisms:a) *P. diminuta* N176, b) *P. diminutaV22*, c) *Pseudomonas sp.* A14; d: *B. laterosporus*

The process of formation of active enzyme from precursor polypeptide involves the following steps [6]:

i. Synthesis of precursor polypeptide as the product of translation of gene,
ii. Transport of precursor polypeptide to periplasmic space and removal of signal peptide, and
iii. Proteolyic cleavage to liberate the active enzymes with two dissimilar subunits.

The components of precursor polypeptide chain of CPC acylase enzyme is presented in Table 4.

Table 4. Components of precursor polypeptides of cephalosporin acylases

Enzyme	Source	No. of amino acid residues				References
		Signal peptide	Alpha subunit	Spacer peptide	Beta subunit	
Cephalosporin-C acylase						
Cephalosporin-C acylase I	*E.coli*[a] JM109	Nil	238	Nil	535	[7]
Cephalosporin-C acylase	*E.coli*[b] JM 109	Nil	238	Nil	535	[7]
GL 7-ACA acylase II	*Pseudomonas* SE 83	Nil	238	Nil	535	[87]
GL 7-ACA acylase						
GL 7-ACA acylase	*Pseudomonas* *GK16*	29	169	Nil	N.A.	[6]
GL7-ACA acylase I	*Pseudomonas* SE 83	Nil	366	Nil	181	[87]
GL 7-ACA acylase I	*E.coli*[c] MC 1061	28	349	Nil	541	[7]
GL 7-ACA acylase I	*E.coli*[d] JM 109	27	607	--	--	[8]

N.A.: Not applicable; Parent organisms: a) *P. diminuta* N176; b) *P. diminuta* V22 c) *Pseudomonas* sp. A14; d: *B. laterosporus*

ESTIMATION OF 7-ACA

Various methods are reported in the literature for estimation of 7-ACA.

A reproducible colorimetric method for the determination of small amounts of 7-ACA was described in 1968 where a yellow colored 7-ACA-ninhydrin chromophore was monitored at 405 nm. The color formation increases uniformly and rapidly, reaching a maximum in 9 minutes and then decreasing [88]. The cephalosporin acylase activity based on the liberation of 7-ACA and 7-ADCA was also measured by the colorimetric method described by Balasingham *et al.* following the formation of a yellow condensation product by reaction of 7-ACA or 7-ADCA with p-dimethyl aminobenzaldyhyde [89].

A spectrophotometric procedure for the assay of cephalosporin-C amidohydrolase activity, based on the determination of 7-ACA produced in the hydrolysis of cephalosporin-C by the enzyme has also been investigated. The same method could also be used as a fluorometric method with a 100 fold greater sensitivity. At pH 4.5, 7-ACA produces a strong fluorophor with florescamine, detectable spectrophotometrically at 378 nm and fluorometrically at an excitation of 378 nm and emission of 495 nm. At this pH the fluorophors formed with cephalosporin-C, proteins and aminoadipic acid show minimal absorbance values [90].

Zhang *et al.* developed a novel method for detecting microorganisms capable of producing cephalosporin-C acylase and/or GL 7-ACA acylase. The method is based on the degradation of 2-nitro-5- (6-bromohexanoyl amino) benzoic acid (NBHAB), a chromogenic substrate, into yellow 2-nitro-5-amino benzoic acid by the action of cephalosporin acylases [65].

High performance liquid chromatography (HPLC) was used for analysis of 7-ACA on an Inertsil ODS-2 column with a mobile phase of phosphate buffer supplemented with 2-4% methanol. The 7-ACA was detected in UV range at 254 nm [1, 7-8].

CONCLUSION

Cephalosporin acylases are industrially important enzymes which hydrolyze cephalosporin-C to 7-aminocephalosporanic acid; required for the production of semisynthetic cephalosporins. The work on single step enzymatic conversion of CPC to 7-ACA is underway in various laboratories, but the yield is low as compared to the two enzymatic process. Though, literatures are available on the modification of β-lactam antibiotics using both wild and recombinant strains for the biosynthesis of 7-ACA, we still need an efficient biotransformation process towards the production of 7-ACA in a single step enzymatic pathway.

ACKNOWLEDGMENT

V. K. Nigam wishes to acknowledge the financial assistance received from Council of Scientific and Industrial Research as a research fellowship (1992-1997) and Department of Science and Technology, Government of India as a post doctoral fellowship (2003-2006) respectively. The authors are also thankful to Dr. R. K. Gothwal, Scientist, BISR, Jaipur for his valuable suggestions during the preparation of this chapter.

REFERENCES

[1] Aramori, I., Fukagawa, M., Iwami, M., Yokota, Y., Kojo, H., Kohsaka, M., Ueda, Y. and Imanaka, H. (1991a). Isolation of soil strains producing new cephalosporin acylases. *J. Ferment. Bioengg.*, 72, (4), 227-231.

[2] Sudhakaran, V. K., Deshpande, B. S., Ambedkar, S. S. and Shewale, J. G. (1992). Molecular aspects of penicillin and cephalosporin acylases. *Process Biochem.*, 27, (3), 131-143.

[3] Shibuya, Y., Matsumoto, K. and Fujii, T. (1981). The isolation and properties of *Pseudomonas* mutants with an enhanced activity of 7-beta-(4-carboxybutanamido) cephalosporanic acid acylase. *Agric. Biol. Chem.*, 45, (7), 1561-1567.

[4] Ichikawa, S., Murai, Y., Yamamoto, S., Shibuya, Y., Fujii, T., Komatsu, K. and Kodaira, R. (1981a). The isolation and properties of *Pseudomonas* mutants with enhanced productivity of 7β-(4-carboxybutanamido) cephalosporanic acid acylase. *Agric. Biol. Chem.*, 45, (10), 2225-2229.

[5] Ichikawa, S., Shibuya, Y., Matsumoto, K., Fujii, T., Komatsu, K. and Kodaira, R. (1981b). Purification and properties of 7β-(4-carboxybutanamido) cephalosporanic acid acylase produced by mutants derived from *Pseudomonas. Agric. Biol. Chem.,* 45, (10), 2231-2236.

[6] Matsuda, A. and Komatsu, K. (1985). Molecular cloning and structure of the gene for 7-beta-(4-carboxybutanamido) cephalosporanic acid acylase from a *Pseudomonas* strain. *J. Bacteriol.*, 163, (3), 1222-1228.

[7] Aramori, I., Fukagawa, M., Tsumura,M., Iwami, M., Isogai, T., Ono, H., Ishitani, Y., Kojo, H., Kohsaka, M., Ueda, Y. and Imanaka, H. (1991b). Cloning and nucleotide sequencing of new glutaryl 7-ACA and cephalosporin C acylase genes from *Pseudomonas* strains. *J. Ferment. Bioengg*., 72, (4), 232-243.

[8] Aramori, I., Fukagawa, M., Tsumura, M., Iwami, M., Ono, H., Kojo, H., Kohsaka, M., Ueda, Y. and Imanaka, H. (1991 c). Cloning and nucleotide sequencing of a novel 7 beta-(4-carboxybutanamido) cephalosporanic acid acylase gene of *Bacillus laterosporus* and its expression in *Escherichia coli* and *Bacillus subtilis. J. Bacteriol.,* 173, (24), 7848-7855.

[9] Monti, D., Carrea, G., Riva, S., Baldaro, E. and Frare, G. (2000). Characterization of an industrial biocatalyst: Immobilized glutaryl 7-ACA acylase. *Biotechnol. Bioengg.,* 70, (2), 239-244.

[10] Seung, W. P., Sang, Y. C., Koo, H. C., Sun, I. H. and Seung, W. K. (2002). New method for the immobilization of glutaryl-7-aminocephalosporanic acid acylase on a modified epoxy/silica gel hybrid. *J. Bioscience Bioengg.,* 94, (3), 218-224.

[11] Tischer, W., Giesecke, U., Lang, G., Roder, A. and Wedekind, F. (1992). Biocatalytic 7-aminocephalosporanic acid production. *Ann. N. Y. Acad. Sci.,* 672, (XI), 502-509.

[12] Kumar, K. K., Sudhakaran, V. K., Deshpande, B. S., Ambedkar, S. S. and Shewale, J. G. (1993). Cephalosporin acylases: Enzyme production, structure and application in the production of 7-ACA. *Hind. Antibiot. Bull.*, 35, 111-125.

[13] Oh, B., Kim, M., Yoon, J., Chung, K., Shin, Y., Lee, D. and Kim, Y. (2003). Deacylation activity of cephalosporin acylase to cephalosporin C is improved by changing the side-chain conformations of active-site residues. *Biochem. Biophys. Res. Commun.,* 310, (1), 19-27.

[14] Pollegioni, L., Lorenzi, S., Roshini, E., Marcone, G. L., Molla, G., Verga, R., Cabri, W. and Pilone, M. (2005). Evolution of an acylase active on cephalosporin-C. *Protein Science*, 14, (12), 3064-3076.

[15] Walksman, S. A., Ekatz, E. and Vining. L. (1958). Nomenclature of the actinomycins. *Proc. Natl. Acad. Sci. U.S.A.,* 44, (6), 602-612.

[16] Martin, J. F. and Demain, A. L. (1980). Control of antibiotic biosynthesis. *Microbiol Mol. Biol. Rev.,* 44, (2), 230-251.

[17] Weinberg, E. D. (1970). Biosynthesis of secondary metabolites: Role of trace metals. *Adv. Microbiol. Physiol.*, 4, 1-44.

[18] Walker, J.B. (1974) Biosynthesis of the monoguanidinated inositol moiety of bluensomycin, a possible evolutionary precursor of streptomycin. *J. Biol. Chem.,* 249, (8), 2397-2404.

[19] Newton, G. G. F. and Abraham, E. P. (1956). Isolation of cephalosporin-C, a penicillin like antibiotic contaning D-alpha-aminoadipic acid. *Biochem. J.,* 62, (4), 651-658.

[20] Demain, A. L. and Zhang, J. (1998). Cephalosporin C production by *Cephalosporium acremonium*: The Methionine Story. *Crit. Rev. Biotechnol.*, 18, (4), 283-294.

[21] Tollnick, C., Seidel, G., Beyer, M. and Schugerl, K. (2003). Investigations on the production of cephalosporin-C by *Acremonium chrysogenum. Advances in Biochemical Engineering /Biotechnology*, 86, 1-45.

[22] Elander, R. P. (2003). Industrial production of beta lactam antibiotics. *Appl. Microbiol. Biotechnol.*, 61, (5-6), 385-392.

[23] Srivastava, P. and Kundu, S. (1999). Studies on cephalosporin-C production in an air lift reactor using different growth modes of *C. acremonium. Process Biochemistry,* 34, (4), 329-333.

[24] Almeida, R. M., Cruz, A. J., Araujo, M. L., Giordano, R. C. and Hokka, C. O. (2001). Modeling and simulation of cephalosporin C production in a fed-batch tower-type bioreactor. *Appl. Biochem. Biotechnol.,* (91-93), 537-549.

[25] Kundu, S., Mahapatra, A. C., Nigam, V. K. and Kundu, K. (2003). Continuous production of cephalosporin-C by immobilized microbial cells using symbiotic mode in a packed bed bioreactor. *Artificial cells Blood substitutes and Biotechnol.*, 31, (3), 313-327.

[26] Cruz, A. J., Pan, T., Giordano, R. C., Araujo M. L. and Hokka, C. O. (2004). Cephalosporin C production by immobilized *Cephalosporium acremonium* cells in a repeated batch tower bioreactor. *Biotechnol. Bioengg.*, 85, (1), 96-102.

[27] Kim, N., Lim, Jung S., Hong, S. and Lim, Seung W. (2005). Optimization of feed conditions in a 2.5-l fed batch culture using rice oil to improve cephalosporin C production by *C. acremonium* M25. *World J. Microbio. Biotechnol.,* 21, (5), 787-789.

[28] Verma, R., Nigam, V. K., Kundu, S. and Ghosh, P. (2006). Morphological differentiation and effect of seed inoculum on the production of CPC by *A. chrysogenum. Asian J. Experimental Sciences*, 20, (2), 275-280.

[29] Nigam, V. K., Verma, R., Kumar, A., Kundu, S. and Ghosh, P. (2007). Influence of medium constituents on the biosynthesis of cephalosporin-C. *Electronic J. Biotechnol.,* (online), http://www.ejbiotechnology.info/content , 10, (2), 230-239.

[30] Wolfe, S., Demain, A. L., Jensen, S. E. and Westlake, D. W. S. (1984). Enzymatic approach to synthesis of unnatural beta lactams. *Science.,* 226, (4681), 1386-1392.

[31] Zhang, Q. J., and Xu, W. X. (1993). Morphological, physiological and enzymatic characteristics of cephalosporin acylase-producing *Arthrobacter* strain 45-8A. *Arch. Microbiol.,* 159, (4), 392-395.

[32] Morin, R. B., Jackson, B. G., Flynn, E. H., and Roeske, R. W. (1969). The reaction of cephalosporin-C with nitrosyl chloride. *J. Amer. Chem. Soc.,* 91, (6), 1396-1400.

[33] Fechtig, B., Peter, H., Bickel, H. and Vischer, E. (1968). Modification of antibiotics [II]. Preparation of 7-aminocephalosporanic acid. *Helv. Chim. Acta.,* 51, (5), 1108-1119.

[34] Frederiksen, R. B. and Emborg, C. (1984). Conversion of cephalosporin-C into 7-phenoxy-acetamido-cephalosporanic acid by acyltransferase of mutants of *Penicillium chrysogenum. Biotechnol. Letters,* 6, (9), 549-554.

[35] Tonkova, A., Ivanova, V., Dobreva, E., Stefanova, M. and Spasova, D. (1994). Thermostable alpha-amylase production by immobilized Bacillus licheniformis cells in agar gel and on acrylonitrile/acrylamide membranes. *Appl. Microbiol. Biotechnol.,* 41, (5), 517-522.

[36] Carvalho, W., Silva, S. S., Vitolo, M., Felipe, M. G. A. and Mancilha, I. (2002). Improvement in xylitol production from sugarcane bagasse hydrolysate achieved by the use of a repeated-batch immobilized cell system. *Z. Naturforsch.,* 57c, (1-2), 109-112.

[37] Srinivasulu, B., Ellaiah, P. and Adinarayan K. (2003). Studies on neomycin production using immobilized cells of *S. marinensis* NUV-5 in various reactor configurations. *AAPS. Pharm. Sci. Tech.*, 4, (3), 369-374.

[38] English, A. R., Huant, H. T. and Sobin, B. A. (1960). 6-Aminopenicillanic acid in urine after oral administration of penicillins. *Proc. Soc. Exp. Biol. Med.,* 104, 405-406.

[39] Batchelor, F. R., Chain, E. B., Richards, M., and Rolinson, G. N. (1961). 6-Aminopenicillanic acid. VI. Formation of 6-aminopenicillanic acid from penicillin by enzymic hydrolysis. *Proc. Roy. Soc. (London). Ser. B.*, 154, 522-531.

[40] Holt, R. J. and Stewart, G. T. (1964). Production of amidase and beta-lactamase by bacteria. *J. Gen. Microbiol.,* 36, 203-213.

[41] Cole, M. (1964). Properties of the penicillin deacylase enzyme of *E. Coli. Nature,* 203, (4944), 519-520.

[42] Cole, M. and Sutherland, R. (1966). The role of penicillin acylase in the resistance of gram-negative bacteria to penicillins. *J. Gen. Microbiol.*, 42, (3), 345-356.

[43] Cole, M. (1966). Formation of 6-aminopenicillanic acid, penicillins, and penicillin acylase by various fungi. *Appl. Microbiol.*, 14, (1), 98-104.

[44] Demain, A. L., Walton, B., Newkirk, J. F. and Miller, I. M. (1963). Microbial degradation of cephalosporin-C. *Nature,* 199, (4896), 909-910.

[45] Ayliffe, G. A. J. (1965). Cephalosporinase and penicillinase activity of Gram negative bacteria. *J. Gen. Microbiol.,* 40, (1), 119-126.

[46] Garber, N. and Friedman, J. (1970). Beta lactamases and resistance of *P. aeruginosa* to various penicillins and cephalosporins. *J. Gen. Microbiol.*, 64, (3), 343-352.

[47] Hamilton Miller, J. M. T., Newton, G. G. F. and Abraham, E. P. (1970). Products of aminolysis and enzymic hydrolysis of cephalosporins. *Biochem. J.*, 116, 371-384.

[48] Richmond, M. H. and Sykes, R. B. (1973). The beta-lactamases of gram-negative bacteria and their possible physiological role. *Adv. Microb. Physiol.*, 9, 31-88.

[49] Matsuda, A., Matsuyama, K., Yamamoto, K., Ichikawa, S. and Komatsu, K. (1987a). Cloning and characterization of the genes for two distinct cephalosporin acylases from a Pseudomonas strain. *J. Bacteriol.,* 169, (12), 5815-5820.

[50] Vandamme, E. J. (1988). In Bioreactors Immobilized Enzymes, and Cells: Fundamentals and Applications (Moo-young, M., ed.), Elsevier Applied Sciences, New York., 261-286.

[51] Reyes, F., Martinez, M. J., Alfonso, C., Copa-Patino, J. L. and Soliveri, J. (1990). Cephalosporin C acylase in the autolysis of filamentous fungi. *J. Pharm. Pharmacol.,* 42, (2), 128-131.

[52] Deshpande, B. S., Ambedkar, S. S., Sudhakaran, V. K. and Shewale, J. G. (1994). Molecular biology of beta lactam acylases. *World J. Microbiol. Biotechnol.,* 10, (2), 129-138.

[53] Nigam, V. K. and Kundu, S. (1997). Effect of organic compounds on the biosynthesis of CPC acyalase. *Ind. J. Microbiol.*, 37, 201-203.

[54] Nigam, V. K. and Kundu, S. (1999). Batch production of 7-ACA by different microorganisms-A comparative study. *Ind. Chem. Engr.*, 41, (1) 5-9.

[55] Kameda, Y., Kimura, Y., Toyoura, E. and Omori, T. (1961). A method for isolating bacteria capable of producing 6-aminopenicillanic acid from benzylpenillin. *Nature,* 191, (4793), 1122-1123.

[56] Walton, R. B. (1964). Search for microorganisms producing cephalosporin-C amidase. *Dev. Indl. Microbiol.,* 5, 349-353.

[57] Kaufmann, W. and Bauner, K. (1964). The production of penicillin amidase by *E. coli.* ATCC 9637. *J. Gen. Microbiol.,* 35, 4.

[58] Daumy, G. O., McColl, A. S. and Apostolakos, D. (1982). Repression of penicillin G acylase of *Proteus rettgeri* by tricarboxylic acid cycle intermediates. *J. Bacteriol.*, 152, (1), 104-110.

[59] Szentirmai, A. (1964). Production of penicillin acylase. *Appl. Microbiol.,* 12, (3), 185-187.

[60] Vandamme, E. J. and Voets, J. P. (1973). Some aspects of the penicillin V-acylase produced by *Rhodotorula glutinis* var. *glutinis*. *Z. Allg. Mikrobiol.,* 13. (8), 701-710.

[61] Huang, H. T., Seto, T. A. and Shull, G. M. (1963). Distribution and substrate specificity of benzylpenicillin acyalse. *Appl. Env. Microbiol.*, 11, (1), 1-6.

[62] Okachi, R., Kato, F., Miyamura, Y. and Nara, T. (1973). Selection of P. melanogenum KY 3987 as a new ampicillin producing bacteria. *Agri. Bio. Chem.*, 37, (8), 1953-1957.

[63] Savidge, T. A. and Cole, M. (1975). Penicillin acylase (bacterial). *Meth. Enzymol.,* 43, 705-721.

[64] Shimizu, M., Masuike, T., Fujita, H., Kimura, K., Okachi, R. and Nara, T. (1975). Search for microorganisms producing cephalosporin acylase and enzymatic synthesis of cephalosporins. *Agric. Biol. Chem.*, 39, (6), 1225-1232.

[65] Zhang, Q. J., Xu, W. X. and Shi, L. (1991). 2-Nitro-5-(6-bromohexanoylamino) benzoic acid test paper method for detecting microorganisms capable of producing cephalosporin acylases. *Anal. Biochem.,* 196, (2), 201-206.

[66] Binder, R. G., Numata, K., Lowe, D. A., Murakami, T. And Brown, J. L. (1993). Isolation and characterization of a *Pseudomonas* strain producing glutaryl 7-ACA acylase. *Appl. Environ. Microbiol.*, 59, (10), 3321-3326.

[67] Lowe, D. A. (1989). Immobilized beta lactam acylases. *Dev. Indl. Microbiol.,* 30, 121-130.

[68] Rolinson, G. N. (1988). The influence of 6-aminopenicillanic acid on antibiotic development. *J. Antimicrobial Agents Chemother.,* 22, 5-14.

[69] Vandamme, E. J. and Voets, J. P. (1974). Microbial penicillin acylases. *Adv. Appl. Microbiol.*, 17, (0), 311-369.

[70] Halim, A. and Sayed, M. M. (1992)."In Handbook of Applied Mycology-Fungal Biotechnology" (eds. D. K. Arora, R. P. Elander, K. G Mukherji, Marcel Dekker Inc., New York.) 4, 517.

[71] Isogai, T., Fukagawa, M., Aramori, I., Iwami, M., Kojo, H., Ono, T., Ueda, Y., Kohsaka, M. and Imanaka, H. (1991). Construction of a 7-aminocephalosporanic acid (7ACA) Biosynthetic Operon and Direct Production of 7ACA in *Acremonium Chrysogenum. Biotechnology.*, 9, (2), 188-191.

[72] Serizawa, N., Nakagawa, K., Haneishi, T., Kaminura, S. and Naito, A. (1980). Enzymatic conversion of cephamycin-C by D-amino acid oxidase from *T. variabilis. J. Antibiotics.*, 33, (6), 585-590.
[73] Kubicek-Pranz, E. M. and Rohr, M. (1985a). Formation of D-amino acid oxidase in the yeast *Trigonopsis variabilis. Can. J. Microbiol.*, 31, (7), 625-628.
[74] Kubicek-Pranz, E. M. and Rohr, M. (1985b). D-amino acid oxidase from *Trigonopsis variabilis:* Comparison of enzyme specificity "in vivo" and "in vitro". *Biotechnol. Letters,* 7, (1), 9-14.
[75] Szwajcer, E. and Mosbach, K. (1985). Isolation and partial characterization of a D-amino acid oxidase active against cephalosporin C from the yeast *T. variabilis. Biotechnol. Letters.*, 7, (1), 1-7.
[76] Chen, J. T., Lin, S. Y. and Tsai, H. (1991). Enzymic and chemical conversions of cephalosporin C to 7-(glutarylamido) cephalosporanic acid. *J. Biotechnol.*, 19, (2-3), 203-210.
[77] Huber, F. M., Vicenzi, J. T. and Tietz, A. J. (1992). High yielding culture conditions for the biosynthesis of D-amino acid oxidase ty Trigonopsis variabilis. *Biotechnol. Letters,* 14, (3), 195-200.
[78] Vicenzi, J. T. and Hansen, G. L. (1993). Enzymatic oxidation of cephalosporin C using whole cells of the yeast *Triginopsis variabilis* within a "cross-flow filter-reactor" *Enz. Microb. Technol.,* 15, (4), 281-285.
[79] Dey, E. S., Flygare, S. and Mosbach, K. (1991). Stabilization of D-amino acid oxidase from yeast Trigonopsis variabilis used for production of glutaryl 7-ACA from cephalosporin-C. *Appl. Biochem. Biotechnol., 27*, (3), 239-250.
[80] Alfani, F, Cantarella, M, Cutarella, N, Gallifuoco, A, Golini, P and Bianchi, D. (1997). Enzymatic conversion of cephalosporin C into glutaryl 7-aminocephalosporanic acid. A study in different reactor configurations. *Biotechnol. Lett.,* 19, (2), 175-178.
[81] Nigam, V. K., Kundu, S. and Ghosh, P. (2005). Single step conversion of CPC to 7-ACA by free and immobilized cells of *P. diminuta. Appl. Biochem. Biotechnol.,* 126, (1), 13-21.
[82] Nigam, V. K. Kundu, S. and Ghosh, P. (2007a). Reusability of entrapped cells of *P. diminuta* for production of 7-ACA. *Appl. Biochem. Biotechnol.,* 141, (1), 119-126.
[83] Nigam, V. K. Kundu, S. and Ghosh, P. (2008). Continuous production of 7-ACA by immobilized cells of *P. diminuta. Biocatal. Biotrans.,* 27, (1), 60-65.
[84] Claridge, C. A., Goureritch, A. and Lein, J. (1960). Bacterial penicillin amidase. *Nature,* 187, (4733), 237-238.
[85] Rolinson, G. N., Batchelor, F. R., Butterworth, D., Cameron-Wood, J., Cole, M., Eustace, G. C., Hart, M. V., Richards, M. and Chain, E. B. (1960). Formation of 6-aminopenicillanic acid from penicillin by enzymatic hydrolysis. *Nature,* 187, (4733), 236-237.
[86] Takahashi, T., Yamazaki, Y., Kato, K. and Isono, M. (1972). Enzymatic synthesis of cephalosporins. *J. Am. Chem. Soci.,* 94, (11), 4035-4037.
[87] Matsuda, A., Toma, K. and Komatsu, K. (1987b). Nucleotide sequences of the genes for two distinct cephalosporin acylases from a *Pseudomonas* strain. *J. Bacteriol.,* 169, (12), 5821-5826.
[88] Marrelli, L.P. (1968). Colorimetric method for determination of 7-ACA and related compounds. *J. Pharm., Sci.,* 57, (12), 2172-2173.

[89] Balasingham, K., Warburton, D., Dunhill, P. and Lilly, M. D. (1972). The isolation and kinetics of penicillin amidase from *Escherichia coli. Biochem. Biophys. Acta.,* 276, (1), 250-256.

[90] Reyes, F., Martinez, M. J. and Soliveri, J. (1989). Determination of cephalosporin-C amidohydrolase activity with fluorescamine. *J. Pharma. Pharmacol.,* 41, (2), 136-137.

In: Recent Trends in Biotechnology, Volume 1
Editors: M.P. Singh, A. Agrawal and B. Sharma
ISBN 978-1-60876-148-7

Chapter 13

THYMIC PEPTIDES: HORMONES WITH BIOLOGICAL RESPONSE MODIFYING POTENTIAL AND MULTI-UTILITY THERAPEUTIC APPLICATION IN LIFE THREATENING PATHOLOGIES

Naveen Kumar Vishvakarma, Pratima Shrivastava and Sukh Mahendra Singh*
School of Biotechnology, Banaras Hindu University,
Varanasi-22 1005,U. P.,India

ABSTRACT

Thymus, a gland responsible for production of T lymphocyte has been demonstrated to be a source of a variety of short peptide hormones collectively designated as thymic peptides. These polypeptides have gained increasing clinical therapeutic applications in a wide variety of patho-physiological anomalies cancer, neuroendocrine disorders, infections, senescence and autoimmune diseases. The thymic peptides depending on their chemical composition have been categorized into different families among which thymosins are the most investigated ones. Thymosin-α and –β have been shown to participate in immunomodulation and tumor progression. We have demonstrated from studies in our laboratory that thymosinα1 can reverse tumor induced immunosuppression for therapeutic benefit of the tumor bearing host. This chapter provides a brief review of current state of knowledge about the biological functions of thymic peptides and their potential for clinical applications and possible future directions for their therapeutic optimization.

INTRODUCTION

Recent emphasis in the field of immunotherapy has been on identifying and characterizing naturally occurring or synthetic small peptides with potential for biological response modifications for achieving therapeutic efficacy in diverse clinical situations. Some of such peptides with immense therapeutic potential are expressed in thymus. These peptide hormones commonly known as thymic peptides, can modulate a variety of immune and non-immune physiological processes. Table 1 presents a summary of the main thymic peptides which have been characterized with respect to their biological functions and put on clinical trials for treatment of a number of pathological conditions.

The major thymic peptides, well characterized for their biological actions, belong to a family designated as thymosins of less than 10 kDa size. Thymosins are of three main types namely: α, β and γ depending on their iso-electric points. Individual thymosins have been further denoted by numerical subscript, where the numerical subscripts denote the order of their characterization and/or isolation. Moreover it has been observed that thymosins are not only elaborated by thymus but also show a wide range of tissue distribution. The thymosins have found tremendous clinical applications for treatment, diagnosis and prognosis of several diseases. It has been reported that thymosin exerts therapeutic benefits in pathological condition such as immunodeficiency, a variety of malignancies, senescence, autoimmune and neuroendocrine disorders and infections (1-3). Various clinical and preclinical evidences point towards a tremendous potential of utilizing the thymosins in designing future therapeutic treatment of life threatening pathological conditions such as malignancies, immuno-deficiencies and acute and chronic infections. Considering such major role of thymosins as highly potent biological response modifiers, in this review the main emphasis has been given on the description of the thymosin family with small notes also appended in relation to the known biological actions of other minor groups of thymic peptides.

Table 1: Main thymic peptides attributed with biological modifier action*

Name of thymic peptide	Chemistry	
	Chemical nature/Number of Amino acid residues	**Approx. Molecular weight**
Thymosin fraction 5	Partially purified extract (Calf)	>40 peptides
Thymic Fraction X	Partially purified extract	Major component 4.2kDa
Thymolymphotropin	Partially purified extract (Bovine)	> 20 Peptides
Thymic Humoral Factor	Partially purified extract, not well characterized	
Thymomodulin	Calf thymus acid lysate derivative	1-10kDa
Thymosin alpha 1	28	3108Da
Thymosin beta 1	74	8451Da
Thymosin beta 4	43	4982Da
Thymosin beta 8	39	~4kDa
Thymosin beta 9	41	~4kDa
Thymulin	09	847Da
Thymopoietin	49	5562Da
Thymostimulin	49	5.5kDA
MB-35	35	3756Da
Thymopentin	5	~600Da

*modified from Millington and Buckinghum,4).

Isolation of Thymic Peptides

The thymic peptides were originally isolated from a partially purified bovine thymic extract, named as thymosin fraction 5 (5). A brief overview of the isolation protocol of the thymic peptides is shown in flow chart 1. In order to obtain the thymosin fraction 5 thymic tissue is subjected to a multi-step extraction procedure involving centrifugation, acetone and ammonium sulphate precipitation followed by fractionation on column chromatography using Sephadex-G 25

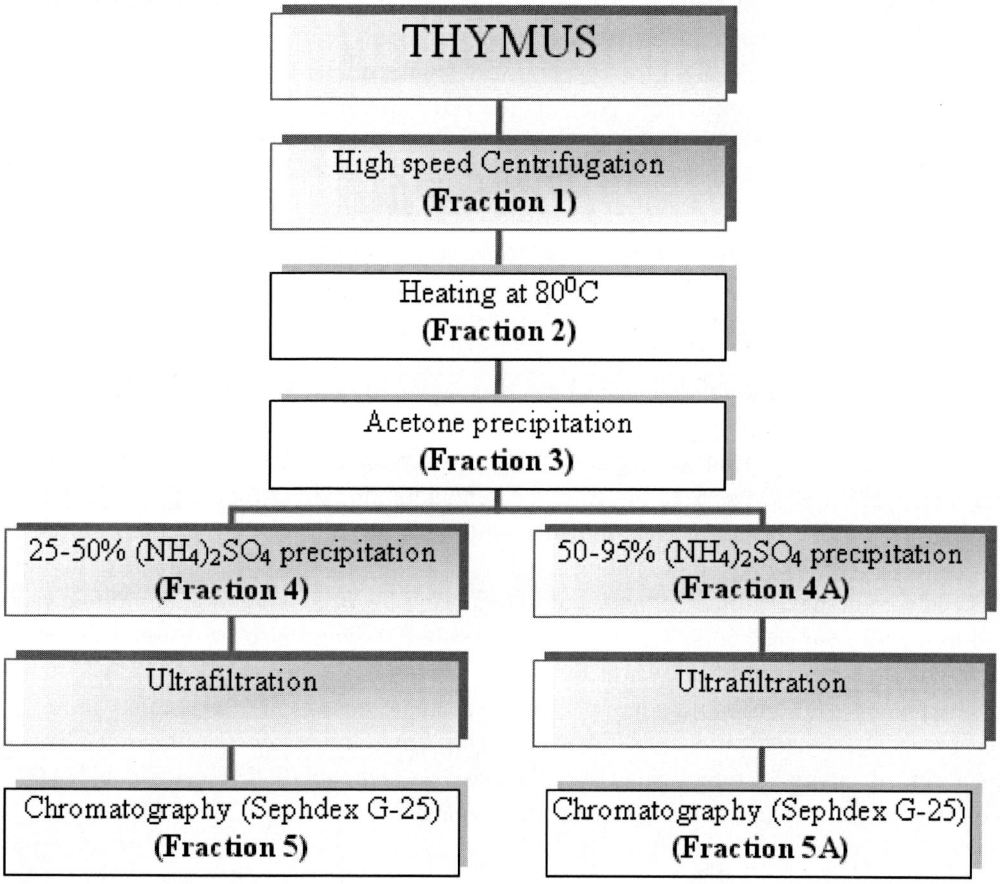

*Adopted and modified from Low *et al*,(7).

Flow Chart 1. Isolation protocol for preparation of thymic peptides*

B THYMOSINS

1. General Biochemistry and Physiology

Approximately 20 thymosins in the β category have been reported to be expressed in a variety of cell types. The β thymosin family consists of highly conserved peptides of approximately 5 kDa size comprising of about 40-44 amino acid residues (8). It is estimated that in mammalian cells, thymosin β4 is the most abundant peptide representing about 70-80% of the total β thymosin content (9). The gene for thymosin β4 has been located on X-chromosome whereas the Y-chromosome has also been reported to have another variant gene for the thymosin β4. The thymosin β4 is 43 amino acid polypeptide (PI 5.1) with a molecular weight of 4.9 kDa. The β4 thymosin functions as an intracellular G actin sequestering peptide and thus serves to provide a large pool of free actin monomers capable of generating new microfilaments. Thus the thymosin β4 through its ability to complex with G actin helps in a rapid reorganization of microfilaments in a variety of cell type needed for specific functions (10-17).

2. β thymosins and Cancer

Overexpression of the β thymosins have been reported in a number of malignancies like: human melanoma, breast, renal, ovarian, uterine, colon and oesophageal carcinoma and small cell lung cancer (9,18). Considering the widespread expression of β thymosins in a variety of cancers, thymosin β10 has been tipped as a potential tumor marker for many carcinomas (9). Similarly thymosin β15 has been shown to influence the metastatic potential of prostate carcinoma cell (9, 19). An additional mechanism of tumor promoting mechanism of thymosin β15 could be derived through its potential to regulate TGF-beta expression in prostate cancer and may influence cell motility (20). Thus thymosin β15 could also be a useful marker for identifying the potential of a given neoplasm for metastasis to secondary sites. Additionally this hormone has also been shown to play a role in tumor metastasis via its ability to regulate plasminogen activator production (21, 22). This again reiterates the important role of thymosin β family in tumorigenesis and metastasis and thus β thymosins could be new molecular targets for anti-tumor therapeutic strategies.

3. β Thymosins in Wound Healing, Angiogenesis, Inflammation, Tissue Reorganization and Cardio-Protective Functions

Accumulating experimental evidences have implicated thymosin β4 in cardiovascular events like angiogenesis, wound healing and apoptosis. Thymosin β4 has been reported to restore ischemic heart (23). Thymosin β4 has also been shown to act as a chemo-attractant for stimulating directional migration of endothelial cell to wounded area (24). In another pioneer work it was indicated that thymosin β4 promotes survival and repair of adult cardiomyocytes through its ability to induce survival kinase AKT (23). Thymosin β4 is also required for coronary vasculogenesis, angiogenesis and arteriogenesis (25). Similarly it has been observed that thymosin β4 induces VEGF expression in lung tumor cell (21). Nevertheless thymosin β4 may also act through its ability to regulate VEGF receptor levels (26). On the same lines thymosin β4 has also been shown to modulate the action of metalloproteinase (27) and thus may be involved in tissue reorganization and organogenesis. Moreover, it was also indicated that thymosin β4 regulates apoptosis (9). Thymosin β4 also reduces lethality by downregulating production of inflammatory mediators in endotoxin induce septic shock (28).

4. β Thymosins and Immune System

Although in general the β thymosins have not been identified as potent immuno-augmentry agents, some sporadic studies strongly indicate their role in the process of hematopoiesis (29), regulation of the functions of granulocytes and mononuclear cells (30) and natural killer (NK) cells (31) and T and B lymphocytes (32, 33). Based on above survey of literature it seems that β thymosins by sequestering G actin, can be used as a potential therapeutic agent to target G actin-dependent cellular molecules involved in pathogenesis of cancer and cardiovascular disorders.

PROTHYMOSIN α

1. Biochemistry and Patho-Physiology

Prothymosinα (ProTα) was initially isolated from rat thymus by Haritos et al (34) as a putative precursor of thymosinα1. Prothymosinα is a highly acidic polypeptide of 109 amino acid size (PI 3.5) and it contains thymosinα1 at its N terminal. Although the thymus is the richest source of ProTα, it has been shown to be ubiquitous expressed in a variety of cell types like kidney, ovary, brain, heart, liver, spleen, muscle, intestine, lung and testis (35-37). It possesses a specific sequence TKKQK at its C terminus, which is responsible for its nuclear localization (38). As a consequence most of the ProTα present in a cell preferentially localizes in the nuclear area (9). The ProTα gene is located on chromosome number 2 and is reciprocally regulated by transcription factors Myc and Max (39). Due to its nuclear localization major activities of ProTα include nucleosome assembly and disassembly, regulation of cell cycle, nucleo-cytoplasmic transport, chromatin decondensation and RNA processing (9, 40).

2. ProTα in Regulation of Immune System and Tumor Progression

ProTα is a biologically active molecule with a variety of immunoregulatory effects on a diverse rget tcell typcs including NK and T lymphocytes (41, 42). Major immunomodulatory activites of ProTα include; augmentation of natural killer cell activity (43), increased antitumor activity of monocytes in melanoma patients (44, 45), enhanced secretion of IL-2 and proliferation of mitogen stimulated peripheral blood mononuclear cells (PBMC) (46, 47) and enhanced response of T-cell proliferative ability to soluble antigens (48). Therefore,

ProTα emerges to have a tremendous immunopharmacological potential in treatment of cancer and other immune system related diseases (36, 49). Treatment of peripheral blood lymphocytes from colorectal tumor patients with ProTα in combination with IL-12 has been shown to cause a partial restoration of lymphocyte deficiencies (50, 51). It has thus been suggested that ProTα may be useful for therapeutic intervention in cancer patients. Evstafiva et al. (52) have describe the ProTα fragmentation to be involved in the regulation of programmed cell death in cancer cells. Thus the level of nuclear ProTα fragmentation plays a decisive role in regulating activation of caspases and thus in tumor progression. Further, ProTα has been demonstrated to play the role of anti-apoptotic factor (9). ProTα is over-expressed in variety of cancer tissues and cell lines indicating that ProTα may have a crucial role to play in process of oncogenesis (53). Moreover, ProTα expression pattern was found to get altered in high grade tumors, thereby, confirming that this peptide may be evolved in the process of tumorigenesis (54). *In situ* hybridization experiments have revealed that ProTα mRNA expression is correlated with the oncogenesis of other cancers like colon cancer (55), liver cancer (56), breast cancer (57), lung cancer (36) and neuroblastoma (58). ProTα through its action in the regulation of gonadal hormones like estrogen has been implicated in manifestation of breast cancer cell proliferation (9). Nevertheless, inhibition of nuclear ProTα expression resulted in consequence inhibition of estrogen induced breast cancer. An additional mechanism of ProTα in regulation cancer progression is derived through its action in regulating expression of tumor suppressor p53 (59).

3. Neuro-Protective Actions and Anti-Ischemic Effects of ProTα

Through its ability to regulate necrosis and apoptosis, ProTα has been demonstrated to play a protective role in the manifestation of retinal ischemia. It has been reported to be up-regulated in ischemia. Thus ProTα produced during ischemic stroke was found to have a defensive role in retinal tissue (60). This indicates that ProTα may have a differential role in regulating proliferation and cell death in normal tissue compared to neoplastic tissues. Another evidence supporting this hypothesis is derived from a recent report of Ueda (61) indicating that ProTα could be considered as a promising target for the regulation of apoptosis in brain tissue and thus may play a central role as 'Cell death mode switch' involved in regulation of brain damage during stroke. Thus ProTα could be a promising target protein for therapeutic manipulation for inducing neuroprotection during stress.

THYMOSINα1 (THYα1)

1. General Introduction and Pathophysiology

Thyα1 is a small peptide, which is one of the most widely, and in the most extensively studied thymic peptide isolated from thymus todate. Thyα1 is a twenty-eight amino acid, 3.1 Kda acidic polypeptide acetylated on its amino terminus. Thyα1 sequence homology exists between all species that have been examined (62). Thyα1 and its mRNA have been localized to the thymic epithelial cells of outer cortex and medulla of thymus (63). Nanogram quantities of thyα1 have also been found in circulation, suggesting endocrine functions for thyα1. It is proposed that thyα1 detected in circulation might be secreted by other non-thymic sources. Localization of thyα1 expression within the thymus seems to be changed with age and in parallel with age associated thymic involution (63-65). Evidences also show that thyα1 is produced in smaller amounts in other tissues including spleen and lymph nodes (66). Moreover, thyα1 expression is reported in non-lymphoid tissues such as lungs, kidneys and brain (28). The immunoaugmentary properties of thyα1 have been well established and have provided novel therapeutic applications in a number of pathological conditions related to infections, malignancy and immunodeficiencies (67, 68). Chemical synthesis of thyα1 has been achieved by both solution phase and solid phase methods. The synthetic preparations have been demonstrated to possess similar biological activities as that of naive form (8).

2. Immunoaugmenting and Hematopoietic Action of Thyα1

Thyα1 has been classified as a potent immunopotentiating agent on the basis of its effect on cytokine secretion, expression of lymphocyte phenotypic markers and lymphocyte functional activity both *in vivo* and *in vitro* (69-74). A summary of its biological activities in modulating immune functions is given in Table 2. These observations provided evidence to indicate that thyα1 played a key role in lymphocyte development. In nude mice thyα1 appears to exert its effect at an early stage of T-cell differentiation and induces a T-cell subpopulation capable of producing IL-3 (75). Thyα1 not only functions in T-cell ontogenesis but also enhances the functions of mature circulating T lymphocytes (76). Thyα1 has been demonstrated to enhance the production and expression of several cytokines including IL-1, IFN-α, IFN-γ on the surface of activated lymphocytes (75- 80). Studies of Serrate et al (81) and Leichtling et al (76) demonstrated the ability of thyα1 to enhance both autologus and allogenic human mixed lymphocyte reaction (MLR) by activation of T-helper cells. Thyα1 can antagonize induction of apoptosis of developing thymocytes (82, 83).

Thyα1 has also been shown to increase the efficiency of antigen presentation by macrophages and it has been suggested that the activation of the macrophage at the time of antigen presentation may be an initial step in the regulation of the immune function by this thymic hormone (84). Moreover, several studies have demonstrated the role of thyα1 in augmenting resistance to infective agents like Herpes simplex virus, Influenza, opportunistic infections and hepatitis B infections (72, 85).

Thyα1 has been suggested to play a role as an autocrine growth modulator for the human breast cancer cell line MCF-7 in culture (86). Other studies have also demonstrated the potential of thyα1 for use as an adjuvant in vaccination, cancer treatment, immune reconstitution, viral infections and age related immunodeficiency problems (87, 88). Thyα1 also augment functional maturation of dendritic cells (DC) and their antitumor activity as demonstrated by us (89) and others in tumors like lymphoma and gastrointestinal carcinoma (90-93). It has been also demonstrated that **thyα1** activates TLR signaling on DC (94). Among other function of DC augmented by thyα1 include priming of DC for anti fungal activity and antiviral activity (93). Nevertheless thyα1 treated DC also regulates Treg activity in vitro (93, 95). Recent studies have demonstrated the role of thyα1 in angiogenesis and tumor regression (96). Garaci et al (2) have shown that thyα1 in combination with cytokines and chemotherapy can be used for treating cancer. Eckert et al (97) has also shown that low molecular weight thyα1 can stimulate the deficient immunocytotoxicity of peripheral blood lymphocytes and monocytes from patients with breast tumor melanoma and colorectal tumor (97). Mayer et al (90) has shown that low molecular weight thymic peptides can stimulate human blood dendritic cells. Similarly, the effect of thyα1 on macrophage derived dendritic cell is unclear. Yang et al (98) demonstrated that thyα1 can enhance the cellular immune functions of elderly patients with malignant tumor (98). Thus thyα1 can act as an adjuvant in promoting and regulating immune responses in condition such as inflammation, immunity and tolerance associated with various pathologies.

In view of the above findings thyα1 has been considered to possess enormous potential for improving immunological disorders associated with various diseases including infection and cancer (68).

3. Drugs based on Thyα1

Thyα1 is able to potentiate the action of cytokines and to reduce the hematological toxicity of cytotoxic drug therapy (72). Interestingly thyα1 has cleared phase III trials for the treatment of hepatitis C and in phase II trials for hepatitis B under the trade name Zadaxin developed by Sci Clone (67, 71, 78, 99-101,). Thus Zadaxin currently approved in around 16 countries as a therapeutic agent in treatment of hepatitis and influenza infections (72). Thyα1 is also marketed under trade name Thymalfasin, which has been proven to be effective against HCV infection (102). Saruc et al (67) have recently suggested a positive role of thyα1 in combination therapy with interferon-γ for negative chronic hepatitis B. Similarly, Arase et al (103) and Andreone et al (78) have evaluated the efficacy of thyα1 therapy in patients with chronic hepatitis B and C (78, 101, 103). Moreover, it has been shown that thyα1 promotes the expression of major histocompatability complex (MHC) class I molecules (104). In addition therapeutic use of thyα1 in treatment of malignant melanoma, hepatocellular carcinoma, drug resistant tuberculosis, AIDS and DiGeorge syndrome has been indicated (105, 106). Till date no cytotoxicity has been reported with thyα1.

4. Thymosin α1 as Transmitters between Neuroendocrine and Immune System

In addition to its immunopotentiating properties evidences also indicate the capability of thyα1 to regulate pituitary gland functions (64, 107). Thyα1 is apparently able to downregulate Thyrotropin stimulating hormone (TSH), Adrenocorticotropin hormone (ACTH) and Prolactin (PRL) secretion in vivo but has been reported to increase blood levels of growth hormone (GH) as well as of nerve growth factor (107, 108). These modulations appeared to occur at least partially; through hypothalamic pathways since production of respective releasing hormone by hypothalamic neurons is also decreased following ***in vitro*** treatment of medial basal α109).

Table 2. Biological properties of Thymosinα

In vitro actions	
1.	Increase in production of interleukin-2, Interferon alpha, interferon gamma and interleukin-7 in activated lymphocytes.
2.	Enhanced expression of high affinity interleukin-2 receptor.
3.	Decreases terminal deoxynucleotidyl transferase (TdT) activity in bone marrow precursors and lymphocytes.
4.	Induction of expression phenotypic T cell Markers.
5.	Enhancement of NK cell recruitment and lytic activity.
6.	Antagonizement of steroid-induced apoptosis in immature thymocytes.
7.	Increase MHC class I expression of lymphoid and non-lymphoid cells.
8.	Increases expression of $CD^{+}4^{+}$ T cells from $CD34^{+}$ stem cells.
In vivo actions	
1.	Increased cytokine production, natural killer activity and antitumor activity in immunosuppresed animals
2.	Enhancement of resistance to many viral, mycobacteria and fungal pathogens in immunosuppresed animals.
3.	Stimulation of viral clearance in woodchuck animal model of hepatitis.
4.	Adjuant effects; restoring immune responses in aged and immunosuppresed animals.
5.	Increases MHC class I expression.
6.	Effective as a monotherapy in enhancing response rates in patients with chronic active hepatitis B.
7.	In combination with interferon increases response rate in patients with haepatitis C
8.	Increases response rate and enhances survival in patients with lung cancer following radiotherapy and in patients with melanoma following chemotherapy.
9.	Increases antibody responses in influenza vaccine in normal elderly volunteers and immunosupressed kidney dialysis patients.

5. Mechanism of Action of Thymosinα1

Expression of receptor(s) for thyα1 has not been proven yet and even the second messenger mechanism(s) remains unclear. Further expression of high affinity receptor for thyα1 has not been detected, although there is some evidence in the literature of its existence (110). Earlier studies of second messenger mechanisms for thymic hormone mediated activity suggested that both cyclic nucleotide and calcium pathways were involved (82). However, thyα1 has been shown to mediate prostaglandin E2 release from immature and splenic lymphocytes in vitro (2, 68). This suggests a role for arachionadate metabolites in the mechanism for thyα1 activity. Baumann et al (82) have demonstrated a role for both cyclic adenosine monophosphate (cAMP) and Protein kinase C (PKC) pathways in the action of thyα1 in antagonizing induction of apoptosis of murine thymocytes. Further, Moody et al (68) have confirmed the role of cyclic AMP in the biological activites of thyα1. These authors also demonstrated that thyα1 stimulates the release of arachidonate metabolites from non-small cell lung carcinoma cells (111). A study using immunofluorescent evidence, demonstrate that thyα1 can bind to murine lymphocytes (2). Thyα1 acts primarily by increasing the efficiency of T cell maturation by increasing the ability of T cells to produce cytokines such as IFN-α, γ, IL-2, IL-3 and IL-6 following antigen or mitogen activation (112-114). Thyα1 also up regulates expression of high affinity IL-12 cytokine receptor (76, 81). A potential mechanism of action for thyα1 is depicted in Figure 1. Moreover, recent reports also indicate that Thyα1 stimulates cell proliferation acting through ERK1/2, JNK and cytokine secretion in pancreatic cancer cells. Thus its action may vary depending on the type of cancer (115).

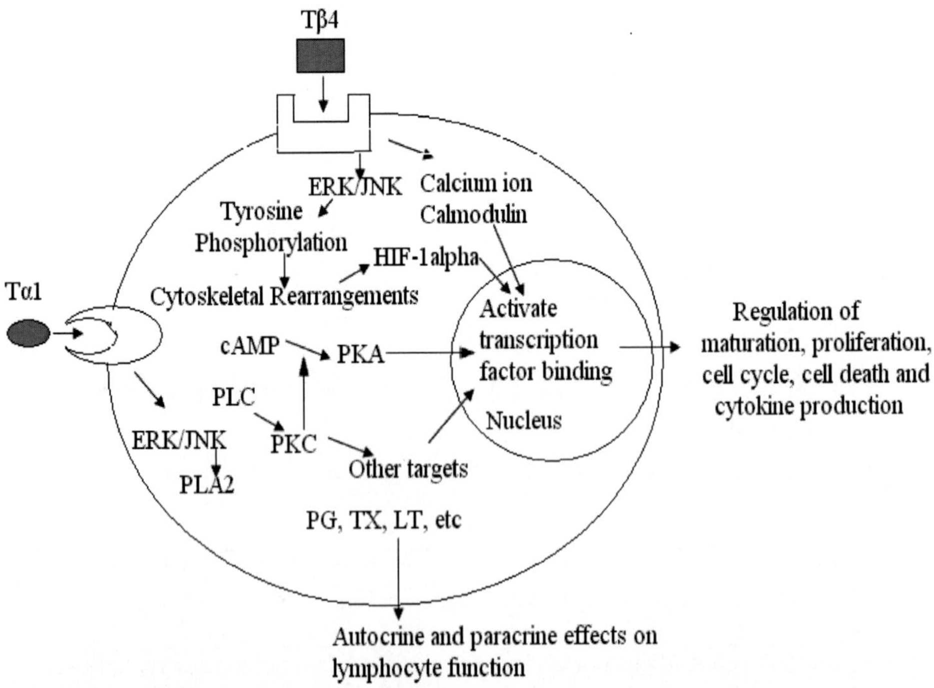

Figure 1. Possible mechanism of action of thymosins.

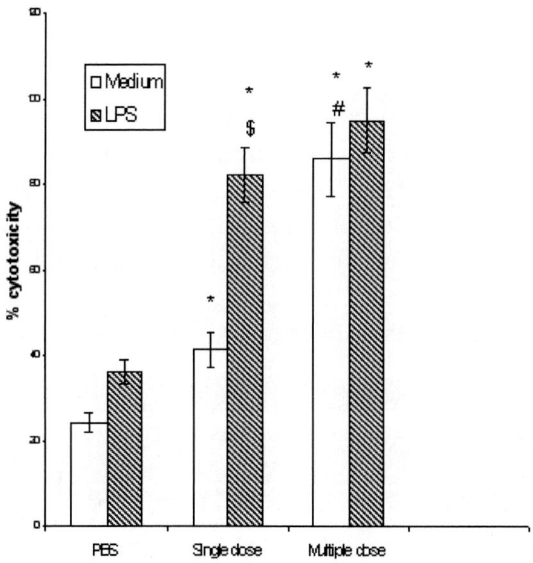

Figure 2. Tumor Asoociated Macrophages (TAM) obtained from mice administered with thyα1 in a single dose (5µg/mouse) administered on day 8th or multiple doses, (5µg/mouse) administered on the 4th, 6th and 8th day of tumor transplantation, were incubated in vitro for 24h in medium alone or containing LPS (10µg/ml). Percentage cytotoxicity against tumor cells was assayed. Values are mean ± SD from a representative experiment done in triplicate. In other experiments similar results were obtained. $*p < 0.05$ vs values of corresponding controls of TAM obtained from mice administered with PBS. $^{\$}p < 0.05$ vs values for corresponding control of TAM incubated without LPS. $^{\#}p < 0.05$ vs values for TAM obtained from DL bearing mice treated with single dose of thyα1 and incubated in vitro in medium alone. (From Shrivastava et al,)(124).

6. Role of Thyα1 in Tumor-Induced Immunosuppression

Our earlier studies with respect to host tumor interaction revealed that thymus undergoes atrophy in a tumor bearing host along with a massive suppression of macrophage tumoricdal activity, cell-mediated and humoral immune responses (116-121). Thymus of the mice bearing a T cell lymphoma regressed depending on the induction of massive cell death in thymocytes, leading to a possible decline in the levels of immuno-potentiating thymic peptides (116-119, 122). Thus it was hypothesized that the decline in the production of thymic peptide could be one of the causes of the inhibited immune responses of the tumor-bearing host. Our studies indicated that in vivo administration of Thyα1 or crude thymic extract could result in the augmentation of the cytotoxicity and accessories function of tumor associated macrophages (89). Our observations further indicated that Thyα1 could alone suffice all the signals required for activation tumor associated macrophages even in the absence of endotoxin signal (89). (Figure 2). Interestingly we also observed that different tissue macrophages displayed heterogeneity in their response to Thyα1 (123) (Table 3). Thus it will be essential to identify the responsiveness of tissue macrophages to thymic peptides prior to their selection for therapeutic application in cancer. Tumor infiltrating Dendritic cell (DC) also plays an important role in host's antitumor response and are as such in an immuno compromised state in a tumor bearing host. Our studies indicated that DC generated from the tumor associated macrophages (TAM) of Thyα1 administered tumor-bearing host (Figure 3) showed an augmented anti tumor activity (124, 125). . This study was the first of its kind to demonstrate that Thyα1 helps not only in the differentiation of Dendritic cell from TAM but also can augment their anti tumor functions (Figure 4). Adoptive transfer of macrophages and TAM derived DC obtained from Thyα1 administered mice resulted in prolongation of the survival of tumor bearing mice (125). Thus these finding demonstrated that thymic peptide can play an important role by their novel action with respect to activation of the myeloid wing of host's antitumor responses. These findings may have long lasting clinical impact in the development of effective immunotherapy using Thyα1 with reference to reversal of immunosuppression associated with malignancies.

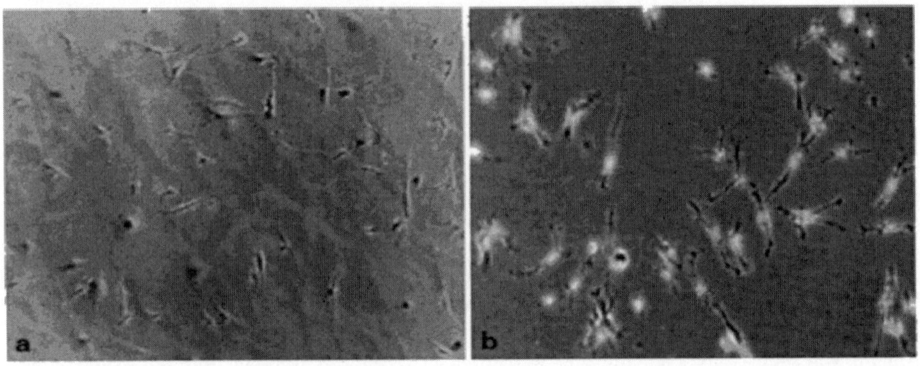

Figure 3. Plate showing the effect of thymosin alpha 1 on the morphology of TAM-derived DC in culture flask. TAM-derived DC obtained from PBS administered (a) and from thymosin alpha1-administered (b) mice. Cytoplasmic spikes are seen in 23% (a) and 60% (b) of cells. From Shrivatava et al. (125).

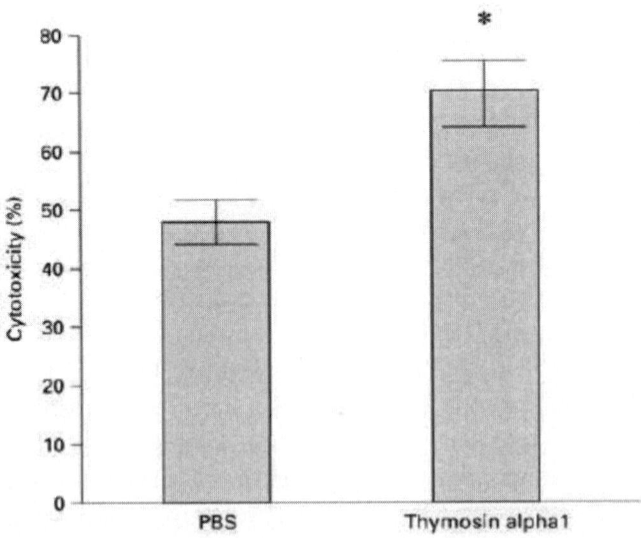

Figure 4. DC obtained from TAM of thyα1- or PBS-administered mice were incubated with DL cells for 24 h for the estimation of cytotoxicity. Values are means + SD of three independent experiments done in triplicate. * $p < 0.05$ vs. values of DC derived from TAM obtained from PBS-administered mice. From Shrivatava et al. (125).

Table 3: Heterogenity in response of tissue macrophages to Thymosin α1 for activation

NO release (μ M/1.5x105 cells)			
Macrophages	**In vitro treatments**	**PBS**	**Thyα1**
Peritoneal Macrophages (PMO)	Medium	15.3±0.12	20.8±2.2*
	LPS	20.7±1.9	25.40±3.5#
Splenic Macrophages (SMO)	Medium	9.2±1.1	17.4±2.6*
LPS		12.4±1.7	20.3±3.0#
Tumor Associated Macrophages (TAM)	Medium	16.2±0.19	26.8±3.5*
LPS		26.2±3.4	34.5±4.4#
Bone Marrow Derived Macrophages	Medium	7.2±1.0	6.2±0.7
(BMDM)	LPS	8.3±0.9	6.5±0.9

PMO, TAM, SMO, and BMDM were obtained from thyα1- or PBS-administered. These macrophages were cultured in vitro for 24 hours in medium alone or containing LPS (10 mg/mL), and the culture supernatant was harvested and analyzed for production of NO. Values are mean±SD from a representative experiment done in triplicate. In other experiments, similar results were obtained. *p<0.05 vs values for corresponding control. #p<0.05 vs. values for corresponding control incubated with LPS. Adapted from Shrivastava et al., (123)

Other Thymic Peptides with Biological Response Modifier Action

Thymulin is a conserved nona-peptide primarily produced by thymic cortical epithelial cells (4). Thymulin has been demonstrated to augment T cell differentiation and the functions of T suppressor cells as well as NK cells. A number of studies also suggest the effect of thymulin on the release of cytokines by T lymphocytes. Therapeutic application of thymulin for correcting T cell abnormalities in AIDS patients and autoimmune disorders has been shown to hold therapeutic promise (111). Thymomodulin which is also component of extract of calf thymus and marketed under trade name Leucotrofina has been administered as a drug for treatment of microbial infections, allergies and immunodeficiencies like HIV infection (126). Thymostimulin or TP-1 is produced for clinical applications by Serono Pharmaceuticals, Italy for HIV infection. TP-1 has been shown to augment T cell activity under immunocompromised situations (127) Thymostimulin has been used in combination with anti-HIV drug AZT to improve the therapeutic efficacy (128). Thymopentin is yet another synthetic derivative of thymic peptide also known as TP-5 and marketed under trade name Timunox for application in AIDS therapy along with AZT (129). Thymic humoral factor (THF) is also a synthetic thymic peptide used for therapeutic applications for restoring cell mediated immune responses (130).

Conclusion

Survey of literature indicates that naturally occurring thymic peptides are highly useful for therapeutic applications in a huge variety of pathologies ranging from cancer, infection, inflammation, cardiovascular, neuroendocrine disorders, autoimmune diseases and senescence. In several countries clinical application of thymic peptides has been approved successfully. Considering the wider role that these peptides play in augmenting antitumor activity of a host against tumor and their direct role in tumor progression, there will be the drugs of choice for clinical applications in near future. Special attention is now needed to understand and explore the hormonal and cytokine network participation of thymic peptides to generate a holistic view of their hierarchy in modulation of biological responses so that optimization of their clinical applications in specific patho-physiological states can be achieved. With innovations in molecular biology, biotechnology and bioinformatics more knowledge about the thymic peptides will be acquired on the basis of which a data base of thymic peptides needs to be created to integrate the available information on potential of their clinical application. These molecules not only have therapeutic applications but can also be utilized for diagnosis as target molecular markers in a variety of pathologies and in their prognosis. Nevertheless, it will also be essential to understand their mode of action of individual thymic peptide on molecular basis more precisely to achieve an understanding of the mechanism for biological response modifying ability with accuracy. Last, but not the least, it is also indispensable to understand the profile of the receptors of thymic peptides and the down stream signal transduction for success in their clinical application. Thymic peptides will no doubt have glorious future and long lasting clinical applications due to the greater role they play in the patho-physiology of several diseases.

Acknowledgment

The authors are grateful to the DBT, Government of India, New Delhi for Financial Support of the work reported in this chapter. CSIR, New Delhi is acknowledged for research fellowship to Naveen Kumar Vishvakarma.

References

[1] Leung, N (2002) Treatment of chronic hepatitis B: case selection and duration of therapy. *J. Gastroenterol. Hepatol.* 17:409-414.

[2] Garaci, E, Pica, F, Sinebaldi-Vallebona, P, and Pierimarchi, P (2003) Thymosin alpha 1 in combination with cytokine and chemotherapy for treatment of cancer. *Int. J. Immunopharcol.* 3:114-1150.

[3] Rasi, G, Pierimarchi, P, Sinebaldi, VP, Colella, F, and Garaci, F (2003) Combination therapy in the treatment of chronic viral hepatitis and prevention of hepatocellular carcinoma. *Int. J. Pharmacol.* 3:1169-1176.

[4] Millington, G, and Buckinghan, JC (1992) Thymic peptides and neuroendocrine-immune communication. *J. Endocrinol.* 133:163-168.

[5] Hannappel, E, and HIff, T (1984) The thymosins, Prothymosin alpha, parathymosin and beta thymosin structure and function. *Vitam. Horm.* 66:257-296.

[6] Goldatein, AL (1994) Clinical application of thymosin aalpha1. *Cancer Invest.* 12:545-549.

[7] Low, TLK, Thurman, GB, Mc-Adoo, M, Mc-Clure, J, Rossio, JL, Naylor, PH, and Goldstein, AL (1979) The chemistry and biology thymosin I. Isolation, characterization and biological activity of thymosin aalpha 1 and polypeptide beta 1. *J. Biol. Chem.* 254:981-986.

[8] Low, TLK, and Goldstein, AL (1984) Thymosins: structure, function and therapeutic application. *Thymus* 6: 37.

[9] Chen, C, Li, M, Yang, H, Chai, H, and Fisher, W (2005) Roles of thymosins in cancer and other organ systems. *World J. Surg.* 29:264-270.

[10] Clauss, IM, Wathelet, MG, Szpirer, J, et al. (1991) Human thymosin b4/6- 26 gene is part of a multigene family composed of seven members located on seven different chromosomes. *Genomics*;9:174–180

[11] Nachmias, VT, Cassimeris, L, Golla, R, et al. (1993) Thymosin b4 (Tb4) in activated platelets. *Eur. J. Cell. Biol.* 61:314–320.

[12] Cassimeris, L, Safer, D, Nachmias, VT, et al. (1992) Thymosin b4 sequesters the majority of G-actin in resting human polymorphonuclear leukocytes. *J. Cell Biol.* 19:1261–1270.

[13] Yamamoto, M, Shoda, A, Minamino, N, et al. (1992) Expression of thymosin b4 gene during Xenopus laevis embryogenesis. *Biochem. Biophys. Res. Commun.* 184:93–99.

[14] Yu, FX, Lin, SC, Morrison-Bogorad, MA, et al. (1993) Thymosin b10 and thymosin b4 are both actin monomer sequestering proteins. *J. Biol. Chem.* 268:502–509.

[15] Heintz, D, Reichert, A, Mihelic, M, et al. (1993) Use of bimanyl actin derivative (TMB-actin) for studying complexation of b-thymosins: inhibition of actin polymerization by thymosin b9. *FEBS Lett.* 329:9–12.

[16] Jean, C, Rieger, K, Blanchoin, L, et al. (1994) Interaction of G-actin with thymosin b4 and its variants thymosin b9 and thymosin b9 met. *J. Muscle Res. Cell Motil.* 15:278–286.

[17] Huff, T, Zerzawy, D, Hannappel, E. (1995) Interactions of b-thymosins, thymosin b4-sulfoxide, and N-terminally truncated thymosin b4 with actin studied by equilibrium centrifugation, chemical cross linking and viscometry. *Eur. J. Biochem.* 230:650-657.

[18] Huff, T, Muller, CSG, Otto, AM, Netzker, R, and Hannappel, E (2001) βThymosin, small acidic peptide with multiple functions. *Int. J. Biochem. Cell Biol.* 32:205-220.

[19] Bao, L, Loda, M, Janmey, PA, et al. (1996), Thymosin beta 15: a novel regulator of tumor cell motility upregulated in metastatic prostate cancer. *Nat. Med.* 2:1322–1328.

[20] Banyard, J, Barrows, C, and Zetter, BR (2009) Differential regulation of thymosin beta 15 isoforms by transforming growth factor beta 1. *Genes Chromosomes Cancer* 48:502-509.

[21] Cha, HJ, Jeong, MJ, and Kleinman, HK (2003) Role of thymosin β4 in tumor metastasis and angiogenesis. *J. Natl. Cancer Inst.* 95:1674-1680.

[22] Al-Nedawi, KN, Czyz, M, Bednarek, R, Szemraj, J and Swiatkowskam, M (2004) Thymosinβ4 induces the synthesis of plasminogen activator1 in cultured endothelial cells and increases its extracellular expression. *Blood* 103:1319-1324.

[23] Bock, MI, Saxena, A, White, MD, Demaio, JM, and Srivastava, D (2004) Thymosin beta 4 activates integrin like kinase and promote cardiac cell migration, survival and cardiac repair. *Nature* 432:466-472.

[24] Malinda, KM, Goldstein AL, and Kleinman, HK (1997) Thymosin beta 4 stimulates directional migration of human umbilical vein endothelial cell. *FASEB J.* 11:474-481.

[25] Smart, N, Rossdeutsch, A, and Reley, PR (2007) Thymosin beta 4 and angiogenesis: mode of action and therapeutic potential. *Angiogenesis* 10:229-241.

[26] Mu, H, Ohashi, R, Yang, H, Wang, X, Li, M Lin, P, Yao, Q, and Chen, C (2006) Thymosin beta 10 inhibit cell migration and capillary like tube formation of human coronary artery endothelial cell. *Cell Motil. Cytoskeleton* 63:222-230.

[27] Blain, EJ, Mason, DJ, Duance, VC (2002) The effect of thymosin beta on artricular chondrocyte matrix mettaloproteinase expression. *Biochem. Soc. Trans.* 30 :779-782.

[28] Badamchian, M, Mora, CA, Baumann, CA, Paino, JE, and Goldstein, AL (1997) Biodistribution of synthetic thymosin alpha 1 in the serum, urine and major organs of the mice. *Int.. J. immunophrmacol.* 19:59-66.

[29] Gondo, H, Kudo, J, White, JW, et al. (1987) Differential expression of the human thymosin-beta 4 gene in lymphocytes, macrophages, and granulocytes. *J. Immunol.* 139:3840–3848.

[30] Hannappel, E, van Kampen, M (1987) Determination of thymosin beta 4 in human blood cells and serum. *J. Chromatogr.*397:279–285.

[31] Lee, H, Yoon, SY, Kang, H, Park, S, Kim, K, Cho, YH, Kim S, Kim, C, Cho, BJ, Lee, WJ, Bang, SI, Park, H, and Cho, D (2009) Thymosin beta 4 inhances NK cell cytotoxicity mediated by ICAM-1. *Immunol. Lett.* 12:72-76.

[32] Pazimino, NH, Ihle, JH, Mc-Evan, RN and Goldstein, AL (1978) Control of thymocyte precursor in bone marrow by thymic hormones. *Cancer Treat. Rep.* 62:1749.

[33] Kannan. L, Rath, NC, Liyanage, R, and Lay, JO (2007) Identification and characterization of thymosin beta 4 in chiken macrophage whole MALDI-TOF. *Ann. N. Y. Acad. Sci.* 1112:425-434.

[34] Haritos, AA, Goodall, GJ, and Horecker, BL (1984) Prothymosin alpha: isolation and properties of major immunoreactive forms of thymosin alpha 1 in rat thymus. *PNAS* 81:1008-1011.

[35] Pineiro, A, Begona, M, Pllar, M, Cordero, OJ, and Nogueira, M (2001) Identification of receptors for prothymosin alpha on human lymphocyte. *Biol. Chem.* 382:1473-1482.

[36] Sasaki, H, Nonaka, M, Fujii, Y, Yamakawa, Y, Fakai, I, Kiriyaama, M and Sasaki, M (2001) Expression of prothymosin alpha gene as a prognostic factor in lung cancer. *Surg. Today* 31:936-938.

[37] Shiau, AL, Lin, PR, Chang, MY and Wu, CL (2001) Retrovirus mediated transfer of prothymosin gene inhibits tumor growth and prolongs survival in murine bladder cancer. *Gene Ther.* 8:1609-1617.

[38] Rubtsob, YP, Zolotukhins, AS, Vorobjeb, IA, et al, (1997) *FEBS Lett.* 413:135-141.

[39] Eilers, M, Sechirm, S, and Bishop, JM (1991) *EMBO J* 10:133-141.

[40] Karetsou, Z, Martic, G, Tavoulari, S, Christoforidis, S, Wilm, M, Gruss, C, and Papamarcaki, T (2004) Prothymosin α associates with the oncoprotein SET and is involved in chromatin decondensation. *FEBS Lett..* 577:496-500.

[41] Baxevanis, CN, Recols, GJ, and Papamichail, M (1993) Prothymosin alpha restore depressed allogenic cell mediated lympholysis and natural killer cell activity in patients with cancer. *Int J Cancer.* 53:264-268.

[42] Eckert, K, Garbin, F, Mourer, HR, Buttner, P, Garbe, C, Szarnecki, J (1995) Prothymosin alpha modulate lymphokine activated killer cell activity and IL-10 production by peripheral blood lymphocyte from melanoma patients in vitro. *Int. J. Immunopharmacol.* 17:555-561.

[43] Grunberg, E, Eckert, K, Mourer, HR, Immenschuh, P, and Kreuser, ED (1997) Prothymosin alpha effects on IL-2 induced expression of LAF-1 on lymphocyte and their adhesion to human umbilical vein endithlial cells. *J. Interferon cytokine Res.* 17:i59-165.

[44] Garbin, F, Eckert, K, Immensehuch, P, Kreusre, ED and Mourer, HR (1997) Prothymosin alpha effects in vitro on the anti tumor activity and cytokine production of blood monocyte from colorectal patients. *Int. J. Immunopharmacol.* 19:323-332.

[45] Grunberg, E, Eckert, K and Mourer, HR (1998) Prothymosin alpha antagonizes the inhibitory effects of TGF beta on the addition of peripheral blood lymphocyte to human umbilical vein endothelial cell. *Int. J. Mol. Med.* 1:741-746.

[46] Cordero, OJ, Sarandeses, CS, Lopez, JL, Cancio, E, Regueiro, BJ, and Nogueira, M (1994) Prothymosin alpha inhances IL-2R expression in normal human T lymphocyte. *Int. J. Immunopharmacol.* 13:1059-1065.

[47] Heidecke, H, Eckert, K, Schutze, F, and Mourer, HR (1997) Prothymosin alpha effects in vitro on chemotaaxis, cytotoxicity and oxidative response of neutrophills from melanoma, colorectal and breast tumor patients. *Int. J. Immunopharmacol.* 19:413-420.

[48] Garbin, F, Eckert, K, Buttner, P, Garbe, C, and Mourer, HR (1994) Prothymosin alpha augments deficient anti tumor activity of monocyte from melanoma patients in vitro. *Anti Cancer Res.* 14:2405-2412.

[49] Pineiro, A, Cordero, OJ, and Nogueira, M (2000) 15 years of prothymosin alpha: contradictory past and new horizons. *Peptides* 21:1433-1446.

[50] Vucovic, D, and Stajnojevic, N (1997) Prothymosin alpha as a biological response modifier. *Int. J. Thymol.* 5:399-408.

[51] Voutsas, IF, Baxevanis, CN, Gritzapis, AD, Missitzis, I, et al (2000) Autologus human carcinoma. *Cancer Immunol. Immunother.* 49:449-458.

[52] Evstafieva, AG, Belov, GA, Kalkum, M, Chichkova, NV, Bogddnov, AA (2000) *FEBS Lett.* 467:150-154.

[53] Letsas, KP, and Frangou, LM (2006) Surfing on prothymosin alpha proliferation and antiapoptotic properties. *Neoplasma* 53:92-96.

[54] Tsai, YS, Jou, YC, Lee, GF, Chen, YC, Shiau, AL, Tsai, HT, Wu, CL, and Tzai, TS, (2008) Aberrant prothymosin-alpha expression in human bladder cancer. *Urology* 73:188-192.

[55] Mori, M, Bernard, GF, Staniunas, RJ, et al. (1993) Prothymosin a mRNA expression correlates with that of c-myc in human colon cancer. *Oncogene* 8:2821–2826.

[56] Wu, CG, Habib, NA, Mitry, RR, et al. (1997) Overexpression of hepatic prothymosin alpha, a novel marker for hepatocellular carcinoma. *Br. J. Cancer* 76:1199–1204.

[57] Orre, RS, Cotter, MA, Subramanian, C, et al. (2001) Prothymosin a functions as a cellular oncoprotein by inducing transformation of rodent fibroblasts in vitro. *J. Biol. Chem.* 276:1794–1799.

[58] Sasaki, H, Sato, Y, Kondo, S, et al. (2001b) Expression of the prothymosin alpha mRNA correlated with that of N-myc in neuroblastoma. *Cancer Lett.* 168:191–195.

[59] Zakharova, NI, Sokolov, VV, Rudkov, VV, Melinikov SV, Vartapetian, AB, Evstafieva, AG (2008) Influence of prothymosin alpha and its mutant on activity of the p53 tumor suppressor. *Mol. Biol. (Mosk.)* 42:673-684.

[60] Fujita, R, Ueda, M, Fujiwara, K and Ueda, H (2009) Prothymosin alpha plays a defensive role in retinal ischemia through necrosis and apoptosis inhibition. *Cell Death Differentiation* 16:349-358.

[61] Ueda, H (2008) prothymosin alpha plays a key role in cell death mode switch, a new concept for neuroprotective mechanisms in stroke. *Naunyn. Sch. Arch. Pharmacol.* 377:315-323.

[62] Low, TLK, Wang, SS, and Goldstein, AL (1983) Solid phase synthesis of thymosin beta 4: Chemical and biological characterization of the synthetic peptide. *Biochem.* 733.

[63] Hirokawa, K, Mc-Clure, JE, and Goldstein,AL (1982) Age related changes in localization of thymosin in human thymus. *Thymus* 4:19.

[64] Goya, RG, Nylor, PH, Goldstein, AL, and Meites, J (1990) Changes in circulating levels of neuroendocrine and thymic hormones during aging in rats: a correlational study. *Exp. Gerentol.* 25:149-157.

[65] Naylor, PH, Friedman-Kein, A, Hersh, E, Erdos, M, and Goldstein, AL (1986) Thymosin alpha1 and thymosin beta 4 in serum: comparison of normal, cord and AIDS serum. *Int. J. Immunopharmacol.* 8:667-670.

[66] Roson, E, Garacia-Caballero, G, Heimer, EP, Felix, AM, andDominguez, F (1990) Cellular distribution of prothymosin alpha and parathymosin in rat thymus and spleen. *J. Histochem. Cytochem.* 38:1889-1894.

[67] Saruc, M, Ozden N, and Yuceyar, H (2003) Thymosin in treatment of HBeAg negative chronic hepatitis B. *Biomed. Sci. Monit.* 9:198-202.

[68] Moody, TW, Tuthill C, Badamchian M, and Goldstein, AL (2002) Thymosin alpha1 inhibits mammary carcinogenesis in fisher rats. *Peptides* 23:1011-1014.

[69] Ahmed A, Smith, AH, Wong, DM, Thurman, GB, Goldstein, AL, and Sell, KW (1978) In vitro induction of lyt surface markers on precursor cells incubated with thymosin polypeptide. *Cancer Treat. Rep.* 62:1739.

[70] Martins, EBMD, and Tuthill, C (2000) Thymosin alpha1, a new biological response modifier for treatment of viral hepatitis and hepatocellular carcinoma. *Rev. Gastroenterol.* Peru 20:58-63.

[71] Ancell, CD, Phipps, J, and Young, L (2001) Thymosin alpha-1. *Am. J. Health Syst. Pharm.* 15:879-885.

[72] Billich A. (2002) Thymosin alpha1.*Current.Opin. Investig. Drugs.* 3:698

[73] Hu SK, Low TL, and Goldstein, AL (1981) Modulation of terminal deoxynucleotide transferase activity by thymosin. *Mol. Cell Biochem.* 41-49.

[74] Goldschneider, I, Ahmed A, Bollum, FJ, and Goldstein, AL (1981) Indunction of terminal deoxynucleotidyl transferase and lyt antigen with thymosin: identification of multiple sub sets of prothymocytes in mouse bone marrow and spleen. *PNAS* 78:2469.

[75] Ohta, Y, Tezuka E, Tamura S, Sugawara M, Nihira S, Imai S, and Yagi, Y (1987) Thymosin alpha1 increases the capability to produce IL-3 but not IL-3 in nu/nu mice. *J. Biol. Response Mod.* 6:181-193.

[76] Leichtling, KD, Serrate SA, and Sztien, MB (1990) Thymosin alpha 1 modulates expression of high affinity IL-2R on normal human lymphocyte. *Int. J. Immunopharmacol.* 2:19.

[77] Hu, SK, Badamchian, M, Mitcho, YL, and Goldstein, AL (1989) Thymosin inhances the production of IL-1 alpha by human peripheral blood monocytes. *Lymphokine Res.* 8:203-214.

[78] Andreone P, Gramenzi A, Cursaro, C, Felline, F, Loggi, E, D'Errioca, A, Spinosa, M, Lorenzini S, Biselli M, Bernardi, M (2004) Thymosin alpha 1 plus interferon alpha for naive patients withchronic hepatitis C: results of randomized controlled pilot trial. *J. Viral Hepat.* 11:69-73.

[79] Favalli, C, Jezzi, T, Mastino, A, Rinalid, Garaci, C, and Garaci E (1985) Modulation of NK cell activity by thymosin alpha 1 and interferon. *Cancer Immunol. Immunother.* 20:189.

[80] Frasca, D, Garavani, M, Doria, G (1982) Recovery of T cell functions in aged mice injected with thymosin alpha1. *Cell Immunol.* 72:384-391.

[81] Serrate, SA, Schulof, RS, Leondanidis, L, Goldstein, AL, Sztein, MB (1987) Modulation of human NK cell cytotoxic activity, lymphokine production and IL-2 receptor expression by thymic hormone. *J. Immunol.* 139:238-2343.

[82] Baumann, C, Badamchian, M, and Goldestein, AL (1997) Preclinical studies of thymosina1 and thymosin b4. In Thymic Peptides in Preclinical and Clinical Medicine and Update; Maurer, H.R., Goldestin, AL, Hager, ED, (Eds.) *W. Zuckschwerdt Verlag: Munchen*, 13– 27.

[83] Osheroff, PL (1981) The effect of thymosin on glucocorticoid receptors in lymphoid cells. *Cell Immunol.* 60:376 .

[84] Tzehoval, E, Szetein, MB, and Goldstein, AL (2000) Thymosin alpha1 and beta 4 potentiate the antigen presenting capacity of macrophages. *Immunopharmacol.* 18:107-113.

[85] Efferos, R (1988) The effect of thymosin alpha1 on immunity ti influenza in aged mice. Aging: *Immunol. Infect. Dis.* 1:31-40.

[86] Oates, KK, Coss, MC (1992) Biochemical and immunohistological identification of thymosin alpha1 in MCF-7 breast cancer cell. Bull. *World Health Organ.* 47:651-658.

[87] Garaci E, Pica F, Rasi, G and Favalli, C (2000) Thymosin alpha 1 in the treatment of cancer from basic research to clinical application. *Int. J. Immunopharmacol.* 22:1067-1073.

[88] Ohmori, H, Kamo, M, Yamakoshi, K, Nitta, MH, Hikida, M, and Kanayama, N (2001) Restoration of immunocyte function by thymosin aalpha1 in cyclophasphamide induced immunodeficient mice. *Immunupharmacol. Immunotoxicol.* 23:75-79.

[89] Shrivastava, P, Singh, SM, and Singh, N (2003) Effect of thymosin alpha 1 on nitric oxide production by tumor associated macrophages. *Neoplasma* 50:47-53.

[90] Mayer, G, Pohlmeyer, K, Caliebe, A, Heimueller, E, Behnke, B, Steimann G, Lange, C, and Beuth, J (2000) Low molecular thymic peptides stimulate human blood dentritic cells. *Anticancer Res.* 5:2873-2883.

[91] Ma, YL, Zheng, Z, Li, BD, Xie, SZ, Li, GX, Yan, QH, and Cai, JH (2007) The antitumor efforts of thymosin alpha1 on tumor lysate pulsed Dendritic cells in colon cancer in vitro and in vivo. *Xi Bao Yu Fen Zi Mian Yi Xue Za Zhi* 23:1046-1049.

[92] Yao, Q, Doan, LX, Zhang, R, Bhardwaz, U, Li, M, and Chen, C (2007) Thymosin alpha1 modulates DC differentiation and functional maturation from human peripheral blood CD4 positive monocytes. *Immunol. Lett.* 110:110-120.
[93] Romani, L, Bistoni, F, Montagonli, C, Gaziano, R, Garaci, E, and Puccetti, P (2007) Thymosin alpha1: An endogenous regulator of inflammation, immunity and tolerance. *Ann. N. Y. Acad. Sci.* 1112:326-338.
[94] Romani, L, et al (2004) Thymosin alpha 1 activates dendritic cell for antifungal Th1 resistance through Toll like receptor signaling. *Blood* 103:4232-4239.
[95] Mosemann, EA, et al (2004) Human plasmacytoid Dendritic cells activated CpG oligonucleotides induce the generation of CD4+CD25+ regulatory T cell. *J Immunol* 173:4433-4442.
[96] Koutrafouri, V, Leondiadis, L, Avoustakis, K, Livaniou, E, Czarnecki, J, Ithakissios, DS, and Evangelatos, GP (2001) Effect of thymosin peptides on the chick coreoallontoic membrane angiogenesis model. *Biochem. Biophys. Acta* 1568:60-66.
[97] Eckert, K, Mayer, G, Stange, R, Calden, M, Karasagakis, K, and Mourer, HR, (1997) Effect of thymic peptide in vitro on the impaired immunocytotoxicity of peripheral blood mononulear cells from tumor patients. *Int. J. Oncol.* 10:481-486.
[98] Yang, S, Liu, ZW, Zhou, WX, and Zhang, YX (2003) Thymosin alpha1 modulates excitatory synaptic transmission in hippocampal neurons in Rats. *Neurosci. Lett.* 350:8-14.
[99] Horwitz, DL (1997) Results of clinical trials of thymosin alpha 1 in chronic hepatitis B conducted in Tiawan, Singapore, Japan and Italy, In : 'thymic peptide in preclinical and clinical medicine :an uptade' Mourer, HR, Goldstein, AL and Hager ED (eds.) *W Zuchschwerdt Verlag Munchen.* 162-167.
[100] Knutsen, AP, Freeman, JJ, Mueller, KK, Roodman, ST, and Bouhasin, JD, (1999) Thymosin alpha1 stimulates maturation of $CD34^+$ Stem cells in an in vitro thymic epithelial organ culture maoldel. *Int. J. Immunopharmacol.* 21:15-26.
[101] Kullavanuaya, P, Treeprasertsuk, S, Thongngam, B, Chaermathi, K, Gonlachanvit, and Suwanagool, P (2001.)The combined treatment of Interferon alpha-2a and thymosinα1 for chronic hepatitis C: The 48 weeks of treatment result. *J. Med. Assoc. Thai.* 84:5462.
[102] Akira, S, Takeda, K, and Kaiso, KT (2001) Toll like receptor: critical protein linking innate and acquired immunity. *Nature Immunity* 2:675-680.
[103] Arase, Y, Tsubota, A, Suzuki, F, Kobayashi, M and Someya, T (2003) A piolot study of thymosin alpha1 therapy for chronic hepatitis patients. *Intern. Med.* 42:941-946.
[104] Giuliani, C, Napolitano, G, Mastino, A, Singer, DS, Cohn, LD, Garaci, E, and Favalli, C (2000) Thymosin alpha1 regulates MHC1 expression in FRTL-5 cells at trancription level. *Eur. J. Immunol.* 30:778-786.
[105] Li, CL, Zhang, T, Sibara, T, Namoto, V, Ono, M, Akisawa, N, Ewasaki, S, Meda, T, and Onishi, S (2002) Thymosin alpha1 accelaeates restoration of T cell mediated neutralizing antibody response in immunocompromised hosts. *Int. J. Immunopharmacol.* 2:39-46.
[106] Chadwick, D, Lopez, PJ, Pires, A, Imarni N, Gotch, F and Villacian, JS (2003) A pilot study of the safety and efficacy of thymosin alpha1 in augmenting immune reconstitution in HIV infected patients with low CD 4 count taking highly active antiretroviral therapy. *Clin. Exp. Immunol.* 134:477-481.

[107] Milenkvoic, L, Lyson, K, Aguila, MC, and McCann SM (1992) effect of thymosin alpha 1 on hypothalamic hormic hormone release. *Neuroendocrinol* 56:674-679.

[108] Savino, W, Postell-Vinay, MC, Smaniotto S, and Dardenn, M (2002) The thymus gland: a target organ for growth hormone. *Scand. J. Immunol.* 55:442-452.

[109] Hall, NR, McGillies, JP, Spangelo, BL, and Goldstein, AL (1985) Evidence that tthymosin and other biological response modifier can function as neuroactive immunotransmitter. *J. Immunol.* 135:806s-811s.

[110] Mazzoccoli, G, Grilli, M, Carughi, S, Urbano, N, Tarquini, R and Perfetto, F (2003) Immune system alterations in lung cancer patients. *Int. J. Immunopath. Pahrmacol.* 16:167-172.

[111] Maurer, HR, Goldstein, AL, and Hager, ED (eds) (1997) In 'thymic peptide I preclinical and clinical medicine: An update'. *W. Zuckschwerdt verlag, Munchen (Publisher)* VII-VIII.

[112] Sztein, MB, and Serrate, SA (1989) Characterization of immunoregulatory properties of thymosin alpha1 on IL-2 production and IL receptor expression on normal human lymphocyte, *Int. J. Immunophamacol.* 11:789-800.

[113] Attia, AW, Badamchian, M, Goldstein, AL, and Spangelo, BL (1993) Thymosins stimulates IL-6 production from rat spleen cell in vitro. Immunopharmacol .26:171-179.

[114] Tizerina, M, Gorosp, WC, Bouman, KL, Badamchian, M, Goldstein AL, Spangelo BL (1997) A novel thymosin peptide stimulate IL-6 release from C6 glioma cell in vitro. *Neuroimmunomodulation* 4:163-170.

[115] Li, M, Feurino, LW, Li, F, Wang, H, Zhai, Q, Fisher, WE, Chen, C and Yao, Q (2007) Thymosin alpha stimulate T cell proliferation by activating ERK1/2, JNK and increasing cytokine secretion in human pancreatic cells. *Cancer Lett.* 248:58-67.

[116] Shanker, A, Singh, SM, and Sodhi, A (2000a) Ascitic growth of a spontaneous T cell lymphoma induces thymic involution 1. Alterations in the repertoire of $CD4^+/CD8^+$ thymocytes. *Tumor. Biol.* 21:288;

[117] Shanker, A, Singh, SM, and Sodhi, A (2000b) Ascitic growth of a spontaneous transplantable T cell lymphoma induces thymic involution 2.Induction of apoptosis in thymocytes. *Tumor Biol.* 21:315.

[118] Shanker, A, Singh, SM, and Sodhi, A (2000c) Impairment of T cell function with the progressive ascitic growth of transplantable T cell lymphoma of spontaneous origin. FEMS Immunol. *Med. Microbiol.* 27:247;

[119] Shanker, A, and Singh, SM (2003) Immunopotentiation in mice bearing a transplantable T-cell lymphoma: role of thymic extract. *Neoplasma*.50: 272-279.

[120] Parajuli, P, and Singh, SM (1996) Alteration in Il-1 and arginase activity of tumor associated macrophages: a role in the promotion of tumor growth. *Cancer Lett.* 107:249-256.

[121] Parajuali, P, and Singh SM (1997) Ascitic growth of a T-cell lymphoma in mice alters the humoral and cellular response to exogenous antigen. *Tumor Biol.* 18:104-112.

[122] Roy, R., Singh, SM, and Shanker, A (2000) Mechanism of thymocyte apoptosis induced by serum of tumor-bearing host: the molecular events involved and their inhibition by thymosinα1. *Int. J. Immunopharmacol.* 22:309-321.

[123] Shrivastava, P, Singh, SM, and Singh, N (2005) Antitumor Activation of Peritoneal Macrophages by Thymosin Alpha-1. *Cancer Investigation* 23:316-322

[124] Shrivastava, P, Singh, SM, and Singh, N (2004a) Activation of tumor associated macrophages by thymosin alpha 1. *Int J Immunopathol Pharmacol* 17:1

[125] Shrivastava, P, Singh, SM, and Singh, N (2004b) Effect of Thymosin Alpha I on the Antitumor Activity of Tumor-Associated Macrophage-Derived Dendritic Cells. *J. Biomed. Sci.* 11:623-63

[126] Valesini, G (1987) A calf thymus acid lysate improves clinical symptoms and T cell defects in early stage of HIV infection: II report. *Eur. J. Cancer Clin. Oncol.* 23:521-531.

[127] Lazerrin, A, et al, (1986) Increase of OK-T4 positive cells during treatment with thymostimulin in parenteral drug addicts with PGL. J. Clin. Lab. Immunol. 20:57-61.

[128] Barcellini, W, et al. (1987) Effect of subcutaneous thymopentin treatment in drug addict with PGL. *Clin. Exp. Immunol.* 67:537-543.

[129] Beall, G (1990) A double blinded placebo-controlled trial of thymostimulin in symptomatic HIV-infective patients. *AIDS* 4:679-681.

[130] Trainin, N, et al (1991) Thymic function. 338:1533.

In: Recent Trends in Biotechnology, Volume 1
Editors: M.P. Singh, A. Agrawal and B. Sharma
ISBN 978-1-60876-148-7

Chapter 14

CYANOBACTERIA: A POTENTIAL BUT NEGLECTED TOOLS FOR BIOTECHNOLOGICAL PROCESSES

Ashwani K Rai[2], S P Tiwari[1], Naveen Sharma[3], Anuradha Rai[1] and Anurag Mishra[1]

[1] Department of Applied Microbiology V.B.S P University, Jaunpur

[2] Department of Botany, Banaras Hindu University, Varanasi 221005, India

[3] P. G. College, Ghazipur (UP), India

ABSTRACT

Cyanobacteria, a small group of photosynthetic prokaryotes cosmopolitan in distribution have long been used as a model system to observe various physiological and biochemical processes. A little attention has so far been paid to the cyanobacterial secondary metabolites. The secondary metabolites not needed for normal growth exhibit antiviral, antibacterial, antifungal, antitumor, herbicidal, algicidal, antimalarial, inflammatory activities and if exploited reasonably can be of immense help to humankind. *Spirulina* is in use as human food because of its high nutritional value, high protein content (70% of dry weight) and presence of all the vitamins. Pigments especially phycobillins are in high demand in cosmetics and as food coloring agent. Cyanobacterial species are efficient catalysts of solar hydrogen production through photolysis of water. They also produce exopolysaccharides and flocculants useful in waste water treatment and clarification of solar ponds. Due to N_2 – fixing ability, cyanobacteria free – living or in symbiosis are often used as biofertilizer and reclamatization of saline usar soil in flooded paddy and sugarcane fields (specially in tropical soils). Traditional varieties of rice acquire their nitrogen need through biological sources such as cyanobacteria. *Azolla-Anabaena* association which reduces acetylene at rates that are 4-8 times greater than those observed for free-living *Anabaena cylindrica* is an efficient green manure. Cyanobacteria having simple growth needs may be exploited for the production of recombinant compounds of medicinal and commercial value as well as for the construction of mutants efficient in N_2-fixation and ammonium liberation. Cyanobacteria, unlike bacteria have been neglected as a source of valuable commercial products. This chapter emphasizes a need of multidisciplinary approach to harness the beneficial aspects related to biotechnological processes.

INTRODUCTION

Cyanobacteria constitute a small class of most diverse group of photosynthetic prokaryotes, containing about 150 genera and 2000 species [29].The oxygen producing photoautotrophs are biochemically highly developed organisms, during evolution undergoing several structural and functional modification, exhibiting versatile physiology and wide ecological tolerance which contribute their competitive success to a wide range of environments. This success is due to their tolerance to high level of UV radiation (due to presence of brown pigment scytonemin which absorb UV light) [9], desiccation tolerance and water stress, capable to photosynthesize at low photon densities most successful in saline environments including soil and competitive advantage of the nitrogen fixing ability. Cyanobacteria have been on earth for over three billion years. During the formation of planet, earth atmosphere was devoid of oxygen and oxygen may have been produced from photolysis of water but, the contribution is considered to be negligible [87]. It is believed that bulk of oxygen came from biological photosynthesis some 3.7 Giga years ago. Evidences available to support the idea through fossil existing, 3 Giga years ago, of organisms identical to present day Cyanobacteria (blue green algae) [88, 89].

The major problem in front of developing countries is environmental clean-up, limited supply of food and energy and crisis of unemployment. When analyzed properly in terms of proper development and economic order, in simplified terms main aspects emerge: population, agriculture, industrial production, energy management and search for new pharmaceuticals. The problem can be overcome by searching or developing new biotechnological processes through suitable approaches. Cyanobacteria may play an important role in this context. The group has been mostly studied for their prokaryotic organization, mechanism of photosynthesis, dinitrogen fixing ability, genetic set up and certain structural aspects such as gas vacuoles and spores. Presently, the significance of cyanobacteria lies primarily to its agronomic importance as a biofertilizer because of efficient nitrogen fixing ability which render them suitable candidate in paddy fields, where nitrogen source is limited or unavailable.

The chemical potential of cyanobacteria lies in producing structurally novel and biologically active natural products has been realized [31], because large numbers of secondary metabolites are produced by them. These compounds exhibit different bioactivities such as toxins, hormones, antibiotics, antifungal and antitumor activity, cytotoxicity to cancer cells, inflammatory, anti-inflammatory activity and neuromuscular toxicity. They can provide valuable tools to basic research.

Even today starvation and malnutrition are wide spread that make sense to find out non-conventional sources of food and protein like *Spirulina*. Cyanobacteria have proved a source of natural chemicals that can substitute synthetic cosmetics. Limited resources of energy, fuels and waste water treatment are the burning problems facing the world. Cyanobacteria may be the answer to all these questions. This chapter deals with the huge potential of cyanobacteria to be used as a tool for the biotechnological researches to generate nutritious compounds, pharmaceuticals products, bioenergy, biofertilizers, flocculants and to overcome agricultural problems.

BIOACTIVE COMPOUNDS

Cyanobacterial strains are capable of producing several toxins simultaneously, only one or two of them are usually dominant in any single strain. Antillatoxin and kalkitoxin, products from the marine cyanobacterium *Lyngbya majuscula* produced concentration-dependent cytotoxicity. Antillatoxin neurotoxicity was produced acutely, whereas kalkitoxin caused a delayed neurotoxic response [7]. Cryptophycins (depsipeptides, peptides with an ester linkage) produced by *Nostoc* sp. GSV 224, have been implicated as promising candidate for anticancer drug [104]. It is a mitotic inhibitor, causing cells to accumulate in mitosis with the disappearance of intracellular microtubules [16]. Tolyporphin, a porphyrin extracted from *Tolypothrix nodosa* exhibited a potent photosenstitising activity against tumor and was five thousand times more effective than the photodynamic treatment (photofrin II) [64]. Curacin A is an antiproliferative (antimitotic) agent found in *L. majuscula* which inhibits the growth of tumor cells [32, 113, 107]. The compound is a potent competitive inhibitor of colchicine binding to tubulin and binds more tightly to tubulin than does colchicine [107]. Broad - spectrum inhibitory metabolite, fischerellin produced by *Fischerella muscicola* (UTEX 1829) was found inhibitory to the growth of eukaryotic algae, cyanobacteria and eubacteria [97]. Symplostatin 1, a new solid tumor selective cytotoxic analogue of dolastatin 10 has been isolated from *Symploca hydanoides* [35]. The linear peptide aeruginosin 103-A, isolated from *Microcystis viridis* (NIES-103) was inhibitory to thrombin [47]. Calothrixin –A and calothrixin-B from *calothrix spp.* are structurally characterized as pentacylic metabolites with an indole (3, 2-j) phenanthridine ring system. They were inhibitory to the growth of a chloroquine – resistant strain of the malaria parasite, *Plasmodium falciparum* and of human HeLa cancer cells [80]. The violet pigment nostocine-A produced by *Nostoc spongiaeforme* TISTR 8169 showed antibiotic, algicidal and herbicidal activities as well as cytostatic activity against human tumor cell lines [38]. Fischerellin-A, the most active allelochemical compound of *Fischerella muscicola* UTEX 1829 is a potent photosystem –II- inhibitor and has significant fungicide and herbicide activity [34]. Isolation of ypaoamide, a new herbivore antifeedant, from the extract of *L. majuscula* is reported by Nagle [65]. Nostocyclamide with antibiotic and algicide activities has been isolated as a major component from Nostoc 31 [102]. Todate, only a segment cyanobacteria have been examined for secondary metabolites and many of the cyanobacterial unidentified bioactive and biomedically interesting compounds remain to be characterized. The very high incidence of novel, biologically active compounds indicates the tremendous potential of cyanobacteria for useful natural products.

Another important aspects is the production of recombinant compounds, as cyanobacteria being autotrophs, have simple growth requirements (light, CO_2 and inorganic elements). The nucleic acids from *Nostoc elliposospurum* encoding cyanovirins may be used in the recombinant production of virucide proteins useful in treatment of virus infections such as human immune deficiency virus [13]. *Synechococcus* PCC 6301 transformed with mosquitocidal toxin genes of *Bacillus thruingiensis* and *Bacillus sphaericus* appeared to be the most suitable organism for use in the natural mosquito habitat, as it could be ingested and digested by mosquito larvae [85]. *Synechococcus* sp. PCC 7942 transformed with pyruvate – decarboxylase and alcohol-dehyrogenase genes of *Zymomonas mobilis,* is able to produce ethanol [26]. Production of ethanol by cyanobacteria is potentially a system for converting solar energy and CO_2 into a valuable product.

Protein and Fine Chemicals

Cyanobacteria are good source of pigments, vitamins, polysaccharides, proteins, pharmaceuticals and other biologically active compounds and may be exploited for production of several chemicals of high market value.

γ- Linolenic Acid (GLA)

Lipid content in cyanobacteria is generally low but *Spirulina* contains 6- 13% lipid on the dry weight basis, half of which is fatty acids [19]. The rare polyunsaturated fatty acid, GLA has medicinal properties and is reported to be 170-fold more effective than linoleic acid, the major constituent of most polyunsaturated oils [41]. The GLA is effective for treatment of atopic eczema [8], alleviating the symptoms of premenstrual syndrome [40], heart disease [46], Parkinson's disease and multiple sclerosis [28]. GLA is precursor for the bodies prostaglandins, master hormones that control many body functions [39]. The prostaglandins E1 are involved in many tasks including regulation of blood pressure, cholesterol synthesis inflammation and cell proliferation [39, 40, 41, 79]. GLA lowers low-density lipoprotein in hypercholesterolemic patient [19]. Ciferri for the first time suggested Spirulina as the source of polyunsaturated fatty acids and the richest source of GLA [18]. The use of inhibitors of fatty acid desaturation can be useful for the over production of GLA.

Another cyanobacterial compound sulphoquinorosyldiacyl glycerol is shown to be effective agent for HIV virus (EC 50 0.1-1μg ml^{-1}) by American National Cancer Institute, but the mechanism for anti-HIV activity is not yet known [78].

Vitamins

Cyanobacteria are a good source of vitamins. *Spirulina,* 20g, is capable to fulfill daily body requirement of vitamin B_{12} and 70% of B_1 (thiamine), 50% of vitamin B_2 (riboflavin) and 12% of B_3 (niacin) [98]. The cyanobcterial strains *Anabaena flos-aquae, Anabaena hassali, Microcystis pulverana, Nostoc punctiforme, Phormidium bijugatum* have been reported to contain these vitamins as well as nicotinic acid and pentothenate [89, 48, 49]. *Spirulina* is also a good source of tocopherol (vitamin E) [42]. Phytol is suitable precursor for the synthesis of several vitamins (A, E, K, K_2).

Pigments

Important pigments of cyanobacteria are chlorophyll a, carotenoids and phycobiliproteins. Chlorophyll a is the most abundant pigment in *Spirulina* (0.8 to 1.5% of dry weight) [66]. Importance of β-carotene is due to its anticancerous property [14]. The highest economic potential proteins are phycobiliproteins used as a natural coloring agent for food and cosmetics, a better substitute of synthetic pigments in drug, food and cosmetic industries as a natural pigment. Dainippon Ink and Chemicals of Japan have commercialized the product with trade name Lina blue. It is mainly used for coloring candy, ice creams, dairy products and soft drinks. The other patent of the company is buffer extract of the blue pigment from *Spirulina* which used in eye shadow, eye liner and lipsticks. Since the product is water insoluble does not fade with water or sweat and does not irritate the skin [21]. A huge demand of these pigments prompted several companies for large commercial investments like 2.5 million by Koor Foods of Israel over last 6 years [113]. Roche product and Wesfarmers sponsored a scientific team in Australia [10].

Carbohydrates

Another important cyanobacterial product is carbohydrate in its both the forms, as storage product and osmotic effectors (e.g. glucosyl glycerol, trihalose etc.).The carbohydrates are generally branched polysaccharides, composed of glucose and structurally similar to glycogen [15]. In *Spirulina*, carbohydrates accounts for 15-22% of dry weight [86]. In *S. platensis* rhamnan is reported to be main polysaccharides. Industrial Science Technology Japan (1997) has patented a new strain of cyanobacterium, *Synechococcus* sp. MA (FERM P-15099), capable to produce poly-β- hydroxybutyrate in high yield, the compound useful in the production of biodegradable thermoplastic resins. Cyanobacterium *Synechococcus* sp. contains upto 0.25M glucosyl glycerol [11]. The level of osmoregulatory carbohydrates can be increased by altering the salinity level of the external growth medium, while the level of others depend upon the nutrient status like nitrogen and phosphorus .

Nutritional Products

French phycologist Dangeard received a sample "Dihe" from a small village market in Republic of Chad (Africa) eaten by the natives [22]. The material was sun dried harden mat of microalgae *Spirulina* collected from the lakes or ponds. It was used by the native population to make sauces. Dihe cakes are in use due to superstitious reasons among the pregnant woman because of the belief that its dark colour will screen the unborn baby from the eye of sorcerers. Léonard and Compére found it to be as dried biscuits around Ford Lamy (Africa) and performed the first chemical analysis reporting very high protein content, 50% of the dry weight [53].

Man's desire to better life quality has added in the demand of the world food products, especially protein rich foods [19]. Cyanobacteria (especially *Spirulina)* because of fast growth rate and high protein content are in use as protein source. Vonshak (1996) [108], has called *Spirulina* as a magic agent due to its property to overcome various problems such as food, cosmetics and anticancerous agent. Due to lack of medical and scientific research, the emphasis is presently centered on human consumption as health food.

Spirulina contains about 70% of dry weight as protein which is higher than the content of other cyanobacterial strains. However, *Spirulina* protein is deficient in methionine, cystein and lysine [1], inferior to meat or milk [18], but superior to plant proteins including legumes. In *Spirulina platensis,* free fatty acid account for 79-80% of the total lipids. The high content of γ-linolenic acid (components of vitamin F) is of considerable human nutritional interest. It has also been reported from other cyanobacteria such as *Synechoccus, Aphanocapsa, Microcystis* and filamentous *Oscillatoria* [45]. Poly β-hydroxybutarate, a carbon reserve and energy source in bacteria is also found in *Spirulina* (6% of dry weight). Carbohydrates accounts for 15-20% of dry weight in *Spirulina platensis* composed only of glucose and glycogen. All the vitamins are found and cyanocobalamin appears to be abundant in *Spirulina platensis* (11mg kg^{-1} of dried cells) [86]. Total nucleic acids content is less than 5% of dry weight, much lower than bacteria and yeast (*Saccharomyces cerevisiae*, the most typical source of single cell protein). Uric acid is regarding toxicology produced in humans and other mammals in the metabolism of purines, and high levels cause pathological conditions (Gout). However, regarding the toxicology experiments on human are a few and incomplete. Salazar has reported that *S. maxima* did not produce adverse effects in mice after subchronic treatment [84]. Feeding of either the oil extracts of *S. maxima* or of its defatted fraction prevented fatty liver development, induced in rats by carbon tetrachloride indicated its hepatoprotective [103]. Pirie reported that administration of 20- 40g of dried *S. maxima* to the athletes for 30-45 days gave good results and similar results were obtained on a child suffering from severe malnutrition [68]. Supply of *Spirulina* to hens slightly stimulated egg production but not the egg size and greatly increased the yolk color. It increases the body weight of young shrimps, prawns and fries of different fishes and in some species sexual maturity reaches earlier, allowing shorter breeding cycle. *Spirulina* supplied fish acquires pink-yellow pigmentation of meat, thus increases its commercial value. Thus, culture of *Spirulina* in clean water may be used to produce biomass for consumption and dirty water to give biomass for feed and extraction of chemicals.

Source of Energy

Molecular hydrogen, a non polluting fuel may play a significant role in the future energy economy, cyanobacterial species are efficient catalysts for solar hydrogen production from photolysis of water [6, 44] where hydrogen evolution is catalyzed by nitrogenase (EC1.18.2.1) via an energy-dependent reaction. A descent review on hydrogen photoproduction by cyanobacteria has been written by Rao and Hall [77]. Mitsui and co-workers at the University of Miami are of opinion that a marine cyanobacterial strain of *Oscillatoria* in a chamber (20 ft sq. × 3ft deep) could produce enough hydrogen to yield 1000 kW hours of electricity per month [61]. Prabaharan and Subramanian have successfully avoided the co-production of oxygen during hydrogen photoproduction by manipulating the growth condition of *Phormidium valderianum* BDU 20041 [70]. The yield was 3-fold higher than the maximum hydrogen production in light. Immobilization of cyanobacteria by entrapment in various matrices (polyurethane and polyvinyl foams) might prove to be an interesting solution to the problem of fuel, energy and chemicals. Markov have patented a method for immobilization of cyanobacteria on the outer surface of fibers in a photoreactor [56]. As aerated aqueous salt solution is pumped, hydrogen produced by cyanobacteria seeps into the in side of the fibers and is removed into a vessel where the solution is heated (80-90° C) to separate the H_2. Hydrogen evolves for 5-6 months at a rate of 2 ml hr^{-1} mg^{-1} cyanobacteria.

Flocculants

Flocculants are used for sedimentation of colloidal substances and cellular materials, and thus applied in a wide range of industrial fields i.e., clarification of tap water, dredging and industrial processes. A variety of fungal and bacterial species produce flocculating substances. Cyanobacteria are also reported to produce exopolysaccharides [69] and flocculants [5], and therefore play an important role in this field. *Phormidium* sp. J-1 produces potent macro-molecular flocculants [5]. Choi reported production of flocculant, macro molecular polysaccharide containing neutral sugars, uronic acid and proteins from *Anabaena* [17]. Binding of cationic dye alcian blue to this flocculant indicate it to be polyanionic. Due to its broad substrate specificity, rapid flocculating activity and thermal stability, it may be exploited for many potential uses such as waste water treatment and clarification of solar ponds etc.

Aquaculture

Aquaculture is a growing industry in the field of food production in developing countries. Being the starting point of aquatic food chain and primary producers, management of algal production is an integral part of many aquaculture operations [4]. Cyanobacteria act as food for molluscs, fishes and crustaceans and together with bacteria they play an important role in maintaining O_2 and CO_2 balance [72].

Waste Water Treatment

The role of cyanobacteria has been reviewed and demonstrated that algal cultures in high-rate oxidation pond for removal of nitrogen and phosphorus and production of algal biomass can be an interesting alternative to testimony waste water treatment [30]. *Spirulina* can also be used in waste water treatment. Cyanobacterial strain *Anabaena* sp. PCC 7120 and *Nostoc ellipsosporum* have the property to degrade organic pollutant lindane [50].

Many pharmaceutical industries produce troublesome wastes. The antibiotic waste impart objectionable odor as well as disturb the valuable microflora of the soil and receiving water bodies. One such is the β–lactam group (i.e., penicillin and ampicillin). The marine cyanobacterium *P. valderianum* BDU (30501) is claimed not only to degrade and detoxify such waste but also uses such antibiotics as its nitrogen source [71]. Recently, Shashirekha found this strain capable to tolerate and grow at a high phenol concentration and removes 38 mg l $^{-1}$ with a retention period of 7 days [90]. This may be useful for treating a variety of phenol-containing industrial waste-waters.

ROLE IN AGRICULTURE

Nitrogen is one of the most important mineral nutrients, commonly limiting crop production and merits as the single largest and most costly input into agricultural productivity. Although plenty of N_2 gas present in the atmospheric plants are unable to take directly from air. Only two groups of organism symbiotic and free living bacteria and cyanobacteria are capable to utilize elemental nitrogen through the process of biological nitrogen-fixation. However, the property is not even universal among these organisms. About 2×10^2 megatons of nitrogen is fixed annually through the process of biological nitrogen-fixation [33]. Paddy field economy is the most suitable (water, temperature and high phosphorus) habitat for the occurrence of the cyanobacteria and constitute highest proportion of total algal biomass. Common cyanobacteria found in rice fields are nitrogen fixers. P. K De for the first time emphasized the importance of N_2-fixing cyanobacteria in rice paddy fields [24]. Specially in tropical soils, traditional rice varieties commonly acquire their nitrogen needs through biological sources specially cyanobacteria.

Usually mixed cyanobacterial cultures of free-living forms are used for algalisation of the paddy fields [105, 82]. Cyanobacteria are manageable system and can be produced in bulk with ease. Soil based cyanobacterial biofertilizers may be produced by the method of Venkataraman [106]. Rice seedlings grown in conjugation with foam-immobilized *A .azollae* strains indicated that the cyanobacteria secreted ammonia into the flood in the paddy field [43]. Another agronomically important plant is free-floating water fern *Azolla* having symbiotic association with the cyanobacterium *Anabaena azollae.* The nitrogen fixed by cyanobacteria reaches the rice plant through (1) release of extracellular nitrogen and (2) recycling as soluble nitrogen which is important for subsequent crop. Mian using pot cultures found that nitrogen from free living cyanobacteria were more readily available to crops in comparison to cyanobacteria living in association [60].

Metting reported that total nitrogen-fixation resulted in 90 Kg N ha^{-1} Year $^{-1}$ [58]. A visible growth usually corresponds to less than 10 Kg N ha^{-1} and a dense bloom may correspond to 10-20 Kg N ha^{-1} [2]. Watanabe reported 30Kg N ha^{-1} per rice crop [112].

The use of cyanobacteria as a biofertilizer is practiced on about 25,000 ha in China where *A .azotica* and *N. sphaeroides* are of common use [59]. In India, approximately two million acres of land inoculated with mixtures of *Aulosira, Anabaena, Nostoc, Plectonema, Tolypothrix* and *Scytonema* (All India Co-ordinated project on Algae, 1979). Cost wise the total expenditure for preparing the inoculum is about 1/3 of chemical fertilizers in terms of ultimate N input in the paddy fields [82]. They found increased grain yield (16%) over control and this has been confirmed in China, Egypt, Philippines and USSR. *A. fertilissima* and *A. doliolum,* alone or in combination with urea significantly increased biomass yield, grain yield and nutritive value of rice [27]. Qualitative as well as quantitative improvement in rice production was maximal with 90 Kg N ha^{-1} together with 12.5 Kg ha^{-1} cyanobacterial biofertilizer [27]. Application of cyanobacterial biofertilizer saved about 25% of the chemical

N demand of the crop. This effect increased every year as reported by Watanabe; to the first crop available of the fixed N is 23-28% [109].

Use of cyanobacteria not only provides fixed N but plays an important role in mobilization of inorganic phosphates by means of extracellular phosphatases and excretion of organic acids [12], improvement of soil structure due to the presence of mucilage that helps in binding soil particles and after death and decay increases humus content of the soil and leads to strong reducing conditions. Pre-soaking of rice seeds with cyanobacterial cultures or extracts enhanced germination of seedling growth. The agents gibberelin, Vitamin B_{12} and amino acids are thought responsible but no consistent report is available.

Green Manuring

Cyanobacterial members form symbiotic association with other group of plants and can confer high rates of nitrogen fixation. Among them, *Azolla-Anabaena* association is an effective green manure for flooded crops and has been in use for centuries in China and Vietnam [62, 55]. It is an out standing green manure because of its high nitrogen-fixation rate and very rapid growth. Its Potential as an alternative or supplemental nitrogen source for rice in tropical as well as temperate regions has been confirmed by many [110, 93, 99]. It doubles its biomass in two days producing biomass of 2.1 ton ha^{-1} [83, 111]. Under ideal conditions it has a potential to fix more than 10 kg N ha^{-1} day^{-1} but in the field condition its capability has been established as 1.1 kg N ha^{-1} day^{-1} [110], sufficient to meet the entire N-requirement of rice crop within a few weeks [54].

Because of the growing concern about conservation of the environment and the need for deploying renewable sustainable resources, the application of *Azolla* as a biofertilizer is of much importance. For the less ecologically damaging agriculture, the process of nitrogen fixation by *Azolla* under suboptimal environmental conditions has been an increasingly pressing issue [76]. Usually three different methods are in practice for the use of *Azolla* as a green manure. *Azolla* is grown either as (a) a monocrop and incorporated into the paddy mud as a basal green manure before the rice is transplanted, or (b) an inter crop with rice and either incorporated as a top dressing manure or allowed to die naturally with out incorporation, and alternatively, (c) as both monocrop and an intercrop.

Symbiotic *Azolla* reduces acetylene at rates that are 4 to 18 times greater than those observed for free living *Anabaena cylindrica* [100]. *A. pinnata* fixes 75 mg N g dry wt^{-1} day^{-1} and produces biomass of 347 tones fresh weight ha^{-1} in a year and contained 868kg N, as much as in 1900 Kg urea [94].

Reclamation of "Usar" Soil

"Usar" means the soil containing low organic matter, N and P. Leather used the term usar for all the saline and alkaline soils and according to him general aridity of land combined with canal irrigation is the main cause in formation of these soils [52]. A vast land area in India is usar and as name suggest they are not fertile, and thus a matter of great concern is to bring them once again under plough. Singh has successfully reclamatized the usar soils through algalization for the cultivation of crops like rice and sugarcane in rotation [95]. According to his observations, during the first shower of rain some group of cyanobacteria (*Cylindrospermum, Scytonema* and *Microcoleus*) starts appearing, later during submerged condition this group detaches and starts floating on the surface and another group appears, the process characteristics of rice field [96]. The process can be practiced twice and thrice in a year. Out of several genera involved in the process two of them *Nostoc commune* and *A .fertilissima* appeared most important. Algal material in the soil conserves organic matter, nitrogen, phosphorus, moisture and consequently converts the sodium clay into calcium type. Although the exact mechanism of conversion is not known, probably through secretion of oxalic acid [96].

CONCLUSION

The toxins and secondary metabolites may prove useful natural compounds from pharmacological points of view. This need interdisciplinary approach to evaluate various unknown properties like antimicrobial property, antiprotozoal activity, antiviral and anticancer property, insecticidal and pesticidal property, antihypertensive agent and muscle relaxant property, and thus the world may find new and important products unknown so for. Most of the antimicrobial agents are cytotoxic. The requirement is to screen them for their toxicity on mammalian cells (especially brain cells, liver cells and nephrons).

Current use of cyanobacteria in rice fields has serious limitations. Cyanobacteria growing normally use much of their fixed nitrogen for their own use, releasing only a small fraction of fixed nitrogen. In fertilized fields, instead of fixing nitrogen, cyanobacteria start using the available nitrogen and become like weeds for the crop. It is obvious that if we produce suitably modified strains capable to compete in field condition, the benefit would increase several fold. The first point is to maximize N_2-fixation or diverting much energy for N_2-fixation. Second, nitrate assimilation, defective mutants have the property to fix N_2 even in the presence of high levels of nitrate [57, 73, 74, 75]. In such strains nitrogenase is not repressed by nitrate.

Glutamine synthatase (GS) defective mutants releasing ammonium continuously would be a better biofertilizer [51], if attempts are made to succeed them in natural habitats like flooded paddy fields. Development of strains with particular features such as resistant to herbicides and insecticides is desirable.

The level of GLA in cyanobacterium was suggested to be regulated by temperature, light intensity, growth phase [20] and nitrogen status of the medium [67]. Studies using the specific inhibitors are necessary for the over production of GLA.

Cyanobacterial released polysaccharides destined to special application especially in biomedical sector, deserve separate consideration, owing to their particularly high added values [23]. Sulphated polysaccharides isolated from *Spirulina platensis* shows anti herpes simplex virus and anti human immunodeficiency virus activities [36, 37]. In this context, the cyanobacterial strain of saline or hypersaline origin where the ability to synthesize and release sulphated exocellular polysaccharide seems to be important [23] and may provide a great resource for the production of new bioactive polymers.

In spite, of huge capability for production of various compounds of nutritional and pharmaceutical importance as well as their role in agriculture, environmental clean up, the group of organism is still neglected one in biotechnological researches. So, a care should be implemented to exploit the potential of such an important group of prokaryotes.

REFERENCES

[1] Aaronson S and Dubinsky Z 1982 Mass production of microalgae; *Experient.* 38 36-40.

[2] Aiyer R S, Sulahudean S and Venkataraman G S 1971 Long term algalization field trial with high yielding rice varieties; *Ind. J. Agric. Sci.* 42 380-383.

[3] All India Co-ordinated project on Algae 1979 Algal bio-fertilizers for rice; pp 61 Indian Agriculture Reasurch Institute New Delhi India.

[4] Bardach J E, Ryther J H and McLarney W O 1972 *Aquaculture: The forming and husbandry of Fresh water and marine organism* (New York: Wiley Interscience).

[5] Bar-Or Y and Shilo M 1987 Characterization of macromolecular flocculants produced by *Phormidium* sp. Strain J 1 and by *Anabaenopsis circularis; Appl. Envion. Microbiol.* 53 2226-2230.

[6] Benemann J R and Weissman J C 1976Biophotolysis: problem and prospects; in *Microbial Energy conversion* pp 413-426 eds Schlegel and Barnea (Germany: Erich Goltze, K G Gottingen).

[7] Berman F W, Gerwick W H and Murray T F 1999 Antillatoxin and kalkitoxin, ichthyotoxins from the tropical cyanobacterium *Lyngbya majuscule*, induce distinct temporal patterns of NMDA receptor- mediated neurotoxicity ; *Toxicon* 37 1645-1648.

[8] Biagi P L, Bordoni A, Masi M, Ricci C, Fanelli C, Patrizi A and Ceccolini 1988 A long team study on the use of evening primrose oil (Efamol) in atopic children; *Drugs. Expt. Clin. Res.* 14 285-290.

[9] Birke L 1974 Marine blue –green algal mats; in *Coastal Ecosystems of the United States,* Vol. 1 pp 331-335 eds H T Odum, B J Copeland, and E A McMahan (Washington DC: The conservation foundation in co- operation with national oceanic and atmospheric administration office of coastal environment).

[10] Borowitzka L J, Borowitzka M A and Moulton T 1984 The mass culture of *Dunaliella salina* for fine chemicals from laboratory to pilot plant; *Hydrobiol.* 116/117 115-134.

[11] Borowitzka L J, Demmerle S, Mackayar M A and Norton R S 1980 Carbon-13 nuclear magnetic resonance study of osmoregulation in a blue green alga; *Sci.* 210 650-651.

[12] Bose P, Nagpal U S, Venkataraman G S and Goel S K 1971 Solubilization of tricalcium Phosphate by blue-green algae; *Curr. Sci.* 40 165-166.

[13] Boyd M R, Shoemaker R H 1999 nucleic acids encoding antiviral cyanovirin proteins isolated from cyanobacteria such as *Nostoc ellipsosporum* useful for treating virus infections (especially human immune deficiency virus); Patent PN 5962668,U.S. Dep. Health-Hum. Serv. Washington, DC, USA.

[14] Burton G W and Ingold K U 1984 β-carotene an unusual type of lipid antioxidant; Sci. 224 569-573.

[15] Casu B, Naggi A and Vercellotti J R 1980 Polisaccaridi di riserva della *Spirulina platensis* estrazione e caratterizzione; in *Prospettive della coltura di Spirulina in Italia* pp. 145-153 ed R Materassi (Rome.Consiglia Nazionale delle Ricerche).

[16] Chen B D, Nakeff A and Valeriote F 1998 Cellular uptake of a novel cytotoxic agent, cryptophtcin-52, by human THP-1 leukaemia cells and H-125 lung tumor cells; *Int. J.Cancer* 77 869-873.

[17] Choi C W, Yoo S A, Oh l-H and Park S H 1998 Characterization of an extracellular flocculating Substance produced by a planktonic cyanobacterium, *Anabaena* sp., *Biotechnol. Lett.* 20 643-646.

[18] Ciferri O 1983 *Spirulina*, the edible micro-organism; *Microbiol. Rev.* 47 551-578.

[19] Cohen Z 1996 The chemicals of *Spirulina*; in *Spirulina platensis (Arthropsira)* pp.175-204 (London: Taylor and Francis Publ.).

[20] Cohen Z, Vonshak A and Richmond A 1987 Fatty acid composition of *Spirulina* strain grown under various environmental conditions; *Phytochem.* 26 2255-2258.

[21] Daianippon Ink and Chemicals Inc. 1980 Production of highly purified alcoholophilic phycocyanin; Japanese Patent 80 77-890.

[22] Dangeard P 1940 Sur une algue bleue alimentarie pour I' homme *Arthrospira platensis* (Nordst) Gomont Actes Soc. Linn. Boreaux Extr. Proces-verbaux 91 39-41.

[23] De Philippis, R., Margheri, M.C., Sili, C. and Vincenzini, M. 1995 Cyanobacteria: a promising group of exopolysaccharide producers. In preceding of 2nd Europian Workshop 'biotechnology of microalgae', pp 78-81 IGV Institut fűr Getreideverrarbeitung GmbH, Bergholz-Rehbrucke.

[24] De P K 1939 The role of blue-green algae in nitrogen fixation in rice field; *Proc. Roy. Soc. London* 127 121-129.

[25] De Pauw N, Morales J and Persoone G 1984 Mass culture of micro algae in aqua culture Systems: progress and constraints; *Hydrobiol.* 116/117 121-134.

[26] Deng M D and Coleman J R 1999 Ethanol synthesis by genetic engineering in cyanobacteria; *Appl. Environ. Microbiol.* 65 523-528.

[27] Dubey A K and Rai A K 1995 Appliaction algal biofertilizers (*Aulosira fertilissimma tenuisI* and *Anabaena doliolum* Bhardwaja) for sustained paddy cultivation in Northern India; *Isr. J. Plant Sci.* 43 41-51.

[28] Dyerberg J 1986 Linoleate- derived polyunsaturated fatty acids and prevention of atherosclerosis, *Nutr. Res.* 44 125-128.

[29] Fott B 1971 Algenkunde, ditaxonomic der einzenmen algenstämme (Stuttgart,Germany: G. Fischer Verlag).

[30] Garbisu C and Hall D O 1993 Removal of phosphate by foam immobilized *Phormidium laminosum* in batch and continuous flow bioreactor; *J. Chem. Biotechnol.* 57 181-189.

[31] Gerwick W H, Mrozek C, Moghaddam M F and Agrawal S K 1989 Novel cytotoxic peptides from the tropical marines cyanobacterium *Hormonthamnion eteromorphyides*; Experiment. 45 115- 121.

[32] Gerwick W H, Proteau P J, Nagle D G, Hamel E, Blokhin A and Slate D L 1994 Structure of Curacin A, a novel antimitotic , antiproliferative and brine shrimp toxic natural products from the marine cyanobacterium *Lyngbya majuscula*, *J. Org. Chem.* 59 1243-1245.

[33] Guerrero M G, Vega J M and Losada M 1981 The assimilatory nitrate reducing system and its regulation; *Ann. Rev. Plant PhysioI.* 32 169-204.

[34] Hagmann L and Juetter F 1996 Fischerellin-A, a novel photosystem-II-inhibiting allelochemical of the cyanobacterium *Fischerella muscicola* with antifungal and herbicidal activity; *Tetrahedron Lett.* 37 6539-6542.

[35] Harrigan G G, Luesch H, Yoshida W Y, Moore R E, Nagle D G, Paul V J, Mooberry S L, Corbett T H and Valeriote F A 1998 Symplostatin 1:A dolastatin 10 analogue from the marine Cyanobacterium *Symploca hytnoides*; J.Nat. Prod. *61* 1075-1077.
[36] Hayashi, T and Hayashi, K (1996) Calcium spirulian, an inhibito of enveloped virus replication from a blue-green algae *Spirulina platensis.* J. Nat. Prod. 59 83-87).
[37] Hayashi, K and Hayashi, T (1996) A natural sulphated polysaccharide, calcium spirulian, isolated from *Spirulina platensis*: in vitro and exo vivo evaluation anti-herpes simplex virus and anti-human immunodeficiency virus activities. AIDS Res. Hum. Retroviruses 12 1463-1471.
[38] Hirata K Takashina J, Nakagami H,Ueyama S, Murakami K, Kanamori T and Miyamoto K 1996 Growth inhibition of various organisms by a violet pigment, nostocine-A produced by *Nostoc spongiaeforme*, *Biosci. Biotechno., Biochem.* 60 1905-1906.
[39] Horrobin D F 1981 The possible roles of prostaglandin E1 and of essential fatty acids in mania depression and alcoholism; in progress *in Lipid Research* Vol 20 539-541ed R T Holman (New York Pergamon Press).
[40] Horrobin D F and Huang Y S 1983 Schizophrenia: the role of abnormal essential fatty acids and prostaglandin metabolism; *Med. Hypoth* 10 329-336.
[41] Huang Y S, Cunnane S C, Horrobin D F and Davignon J 1982 Most biological effects of Zn deficiency corrected by γ-linolenic acid 18:3 ω6 but not by 18:3 ω6; *Antherosclerosis* 41 193-208.
[42] Hudson B Z F and Karis G 1974 the lipids of the alga *Spirulina; J. Sci. Food Agr.* 25 759- 763.
[43] Kannaiyan S, Aruna S J, Kumari S M P and Hall D O 1997 Immobilized cyanobacteria as a biofertilizer for rice crops; *J. Appl. Phycol. 9* 167-174.
[44] Karube I, lkemoto H, Kajiwawa K, Tamiya E and Matsuoka H 1986 Photochemical energy conversion using immobilised blue green algae; *J. Biotechnol,* 4 73-80.
[45] Kenyon C N Rippka R and Stainer R Y 1972 Fatty acid composition and physiological properties of some filamentous blue- green algae; *Arch. Microbiol.* 83 216-236.
[46] Kernoff P B A, Willi A L, Stone K J and McNicol G P 1977 Antithrombotic potential of dihomo γ-linolenic acid in man; Brit. Med. J. 2 1441-1444.
[47] Kodani S, Ishida K and Murakami M 1998 Aeruginosin 103-A , a thrombin inhibitor from the cyanobacterium *Microcystis viridis; J. Nat. Prod.* 61 1046-1048.
[48] Koptera Z P 1970a Biosynthesis of thiamin, riboflavin and vitamin B_{12} by some blue–green algae; *Microbiol Zhournal (Kiev)* 32 555-560.
[49] Koptera Z P 1970b Biosynthesis of biotin, pyredoxin nicotinic acid and pantothenic acids by some blue – green algae; *Microbiol. Zhournal (Kiev)* 32 555-560.
[50] Kuritz T and Wolk C P 1994 Use of filamentous cyanobacteria for biodegradation of organic pollutants; *Envioron. Microbiol.* 61 234-238.
[51] Lattore C, Lee J H, Spiller H and Shanmugan K T 1986 Ammonium ion-excreting cyan bacterial mutants as a source of nitrogen for growth of rice; a feasible study; *Biotechnol. Lett.* 8 507-512.
[52] Leather J W 1897 Agric. Ledger no. 7.
[53] Léonard J and compére P 1967 *Spirulina platensis* (Gom.) Geitl. Algue bleue de grande valeur alimentaire par sa richesse on proteins; *Bull. Jord. Bot. Nat. Bleg.* 37 1-23.

[54] Lumpkin T A 1987 Collection, maintenance and cultivation of *Azolla*; in symbiotic Nitrogen Fixation Technology pp 55-94 ed GH Elkan (New York: Marcel Dekker Inc.).
[55] Lumpkin T A and Plucknett C 1982 *Azolla* as a green manure: use and management in crop production; pp 230 (Boulder Colo: Westview press).
[56] Markov S A 1997 Production of hydrogen from blue-green algae; Russian Patent RU 2083481.
[57] Martin-Nieto J, Flores E and Herrero A 1990 Mutant of *Anabeana variabilis* requiring high level of molybdate for nitrate reductase and nitrogenase activity; *FEMS Microbiol. Lett.* 67 1-4.
[58] Metting B 1981 Systematics and ecology of soil algae; *Bot. Rev.* 147 195-312.
[59] Metting B 1988 Microalgae in agriculture; in *Microalgal biotechnology* pp 288-304 eds ma Borowitzka and L J Borowitzka (Cambridge: Cambridge University Press).
[60] Mian M H 1984 A N-tracer study to differentiate nitrogen supply to flooded rice plants by *Azolla* and *Anabaena* during early and later stages of decomposition; *Ind. J. Agr. Sci.* 54 733-738.
[61] Mitsui A 1980 Saltwater based biological solar energy conversion for fuel, chemical, fertilizer, food and feed; in *Proc. Bio-energy* '80 "pp 486-491 (Washington D C; Bio-energy council).
[62] Moore A W 1969 Azolla; biology and agronomic significance; *Bot. Rev.* 35 17-34.
[63] Moore R E 1984 Public health and marine toxins from blue-green algae; in *Seafood Toxins* pp 369-375 ed E P Ragelis (Washington; Amer. Chem. Soc.).
[64] Morliere P, Maziere J C, Santus R, Smith C D, Prinsep M R, Stobbe C C, Fenning Mc, Goleberg J L and Chapman J D 1998 Tolyporphin: A natural products from cyanobacteria with potent photosensitizing activity against tumor cells in vitro and in vivo. *Cancer Res.* 58 3571-3578
[65] Nagle D G, Paul V J and Roberts M A 1996 Ypaoamide, a new broadly acting feeding deterrent from the marine cyanobactrium *Lyngbya majuscule*; *Tetrahedron Lett.* 37 6263-6266.
[66] Paoletti C, Vincenzini M, Bocci F and Materassi R 1980 Composizione biochimica generale delle biomass e di *Spirulina platensis* , *S. maxima*; in *Prospettive della coltura di Spirilina in Italia* pp.111-125 ed R Materassi (Rome:Consiglio Nazionale delle Ricerche).
[67] Piorreck M, Baasch K-H and Pohl P 1984 Biomass production, total protein chlorophylls, lipids and fatty acids of fresh water green and blue-green algae under different nitrogen regimes; *Phytochem.*23 207-216.
[68] Pirie N W 1975 The *Spirulina* algae; in *Food protein sources* pp.33-34 ed N W Pirie (Cambridge: Int. Biol. Prog. Cambridge Univ. Press).
[69] Plude J L, Parker D L, Schommer O J. Timmerman R J, Hagstrom S A, Joers J J M and Hnasko R 1991 Chemical characterization of polysaccharides from the slime layer of the cyanobacterium *Microcystis flos-aquae* C 3-40 *Appl. Environ. Microbiol.* 57 1696-1700.
[70] Prabaharan D and Subramanian G 1996 Oxygen-free hydrogen production by the marine cyanobacterium *Phormidium valderianum* BDU 20041; *Bioresource Technol.* 57 111-116.

[71] Prabaharan D, Sumathi M and Subramaniam G 1994 Ability to use ampicillin as a nitrogen source by the marine cyanobacterium Phormidium *valderanum* BDU 30501, *Curr. Microbiol.* 28 314-320.

[72] Pruder G 1983 Biological control of gas exchange in intensive aquatic production system; *IEEE* 1002-1004.

[73] Rai A K and Tiwari S P 1999a NO_3^- nutrition and salt tolerance in the cyanobacterium *Anabaena* sp. PCC 7120 and mutant strains; *J. Appl. Microbiol.* 86 991-998.

[74] Rai A K and Tiwari S P 1999b Mutants of the cyanobacterium *Anabaena* sp. PCC7120 altered in nitrate transport and reduction ; *Curr. Microbiol.* 39 237-243.

[75] Rai A K and Tiwari S P 1999c Response to NaCl of nitrate assimilation and nitrogenase activity of the cyanobacterium *Anabaena* sp. PCC7120 and its mutants; *J. Appl. Microbiol.* 87 877-883.

[76] Rai V and Rai A K 1999 Growth behaviour of *Azolla pinnata* at various salinity levels and induction of high salt tolerance; *Plant and Soil* 206 79-84.

[77] Rao K K and Hall D O 1996 Hydrogen production by cyanobacteria; potential, problems and prospects; *J. Mar. Biotechnol.* 4 10-15.

[78] Reshef V Mizrachi E, Maretzki T , Silberstein C, Loya S, Hizi A and Carmeli S 1997 New acylated sulfoglycolipids and digalactolipids and releated known glycolipids from cyanobacteria with a potential to inhibit to reverse transcriptase of HIV-1; J. *Nat. Prod.* 60 1251-1260.

[79] Richmond A 1988 *Spirulina*; in Microalgal biotechnology pp 85-121 eds M A Borowitzka and L J Borowitzka (Cambridge: Cambridge University Press).

[80] Rickards R W, Rothschild J M, Willis A C, de Chazal N M, Kirk J, Saliba K J and Smith G D 1999 Calothrixins- A and-B, novel pentacyclic metabolites from *Calothyix* sp. Cyanobacteria with potent activity against malaria parasites and human cancer cells; *Tetrahedron* 55 13513-13520.

[81] Robbins W J Hervey A and Stebbins M 1951 Further observations on Euglena and B_{12}; *Bullet. of the Torrey Bot. Club* 86 367-373.

[82] Roger P A and Kulasooriya S A 1980 Blue-green algae and rice; pp112, (Los Benos Philippins; *IRRI*).

[83] Roger P A and Watanabe I 1984 Algae and aquatic weeds as a source of organic matter and plant nutrition for wetland rice. In *Organic Matter and Rice* pp 1472168 (Los Benos Philippins; *IRRI*).

[84] Salazar M, Martinez E, Ruiz L E and Chamorro G A 1998 Subchronic toxicity study in mice fed *Spirulina maxima*; *J. Ethnopharmacol.* 62 235-241.

[85] Sangthongpitag K, Delaney S F and Rogers P L 1996 Evaluation of four fresh-water unicellular cyanobacteria as potential hosts for mosquitocidal toxins; *Biotechnol. Lett.*18 175-180.

[86] Santillan C 1982 Mass production of *Spirulina; Experientia* 38 40-43.

[87] Schidlowski M. 1976 Archaean atmosphere and evolution of the terrestrial oxygen budget; in *The early history of the earth* pp 525-535 ed B Windley pp 525-535 (New York: John Wiley & Sons. Inc.).

[88] Schopf J W 1972 Exobiology; in *Frontiers of biology* E L 23 eds A Neuberger and Tatum (Amsterdam: North Holland Publishing Co.).

[89] Schopf J W 1974 Paleobiology of the Precambrian. The age of blue –green algae; *Evol.Biol.* 7 1- 43.

[90] Shashirekha S, Uma L and Subramanian G 1997 Phenol degradation by the marine cyanobacterium *Phormidium valderianum* BDU 30501; *J. Ind.Microbiol. Biotechnol.*19 130-133.

[91] Shilo M 1989 The unique characteristic of benthic cyanobacteria; in *Microbiol. Meth.; Physiological Ecology of benthic microbial communities* pp 207- 213 eds Y Cohen and E Rosenberg (Washington D C; Am. Soc. For Microbiol.).

[92] Singh I P, Miligan K E and Gerwick W H 1999 Tanikolide, a toxic and antifungal lactone from the marine cyanobacterium *Lyngbya majuscule; J. Nat. Prod.* 62 1333-1335.

[93] Singh P K 1977 Multiplication and utilization of fern *Azolla* containing nitrogen fixing algal symbiont as green manure in rice cultivation; *RISO* 26 125-136.

[94] Singh P K 1988 Biofertilization of rice crop; in *Biofertilizers; Potentialities and Problems* pp 109-114 eds S P Sen and P Palit (Kolkata; Plant Physiology Forum).

[95] Singh R N 1950 Reclamation of usar soils in India through blue-green algae; *Nature* 165 325-326.

[96] Singh R N 1961 Role of Blue-green algae in nitrogen economy of Indian agriculture (New Delhi; ICAR).

[97] Srivastava V C, Manderson G J and Bhamidimarri R 1999 inhibitory metabolites production by the Cyanobacterium *Fischerella muscicola; Microviol. Res.* 153 309-317.

[98] Switzer L 1981 *Spirulina* the whole food revolution (Toronto;Proteus Corporation Bantam Books).

[99] Talley S N, Talley B J and Rains D W 1977 Nitrogen fixation by *Azolla* in rice fields; in *Genetic Engg. for Nitrogen –fixation* pp 363-384 eds J M Lyons, R C Valentine, D A Phillps, D W Rains and R C Huffakaer (New York; Plenum Press).

[100] Tang P S, Ding-ji S and Changzheng H 1981 Regulation of energy metabolism (photosynthesis and nitrogen fixation) in blue-green algae; Proc. Joint China-US Phycology Symp. pp339.

[101] Taub F B 1970 Algal culture as a source of food; *Proc. Of the World Maricult.Soc.* 1 110-117.

[102] Todorova A K, Juettner F, Linden A, Pluess T and von Philipsbom W 1995 Nostocyclamide a new macrocylic, thiazole-containing allelochemical from *Nostoc* sp. 31 (cyanobacteria); *J. Org. Chem.* 60 7891-7895.

[103] Torres –Duran P V, Miranda-Zamora R, Paredes –Carbajal M C, Mascher D, Ble-Castillo J ;Diaz- Zagoya J C and Juarez-Oropeza M A 1999 Studies on the preventive effect of *Spirulina maxima* on fatty liver development induced by carbon tetrachloride, in the rat; *J. Ethnopharmacol.* 64 141-147.

[104] Trimurtulu G, Ogino J, Helzel C E, Husebo T L, Jensen C M, Larsen L K, Patterson G M L, Moore R E, Mooberry S L, Corbett T H and Valeriote F A 1995 structure determination conformational analysis chemical stability studies and anti tumor evaluation of the cryptophycins. Isolation of 18 new analogues from Nostoc sp. Strain GSV 224; *J. Am. Chem. Soc.*117 12030- 12049.

[105] Venkataraman G S 1972 Algal biofertilizer and rice cultivation today and tomorrow (New Delhi; Printers and Publishers).

[106] Venkataraman G S 1981 Algal biofertilizer for rice (New Delhi; IARI)

[107] Verdier – Pinard P, Sitachitta N, Rossi J V, Sackett D L, Gerwick W H and Hamel E 1999 Biosynthesis of radiolabeled curacin A and its rapid and apparently irreversible binding to the colchicine site of tubulin ; *Arch. Biochem. Biophys.* 370 51-58.

[108] Vonshak A 1996 Use of *Spirulina* biomass; in *Spirulina platensis (Arthropsira)* pp 205-212 ed A Vonshak (London; Taylor and Francis Publ.).

[109] Watanabe I 1962 Effect of nitrogen-fixing blue-green algae *Tolypthrix tenuis* on the nitrogenous fertility of paddy soil and on the crop yield of rice plants; *J. Gen. Appl. Microbiol.*8 85-91.

[110] Watanabe I 1977 *Azolla* utilization in rice culture; *Int. Rice Res. News Lett.* 2 3-8.

[111] Watanabe I 1987 Summary reports of the *Azolla* programme on the international network on soil fertility and fertilizer evaluation for rice; in IRRI Manila pp197-205.

[112] Watanabe I, Lee K K, Alimagno B V, Sato M, Del Rosario D C and De Guzman M R 1977 Biological N_2 fixation in paddy field studies by *in situ* acetylene reduction assays; *Int. Rice Res. Paper Ser.* 3 1-16.

[113] Weiner J 1985 marine biotech in the Negev desert; *Biotechnol.* 3 41-47.

[114] White J D, kim T-S Nambu M 1997 Absolute configuration and total synthesis of (+)-Curacin A, an antiproliferative agent from the cyanobacterium *Lyngbya majuscule; J. Am. Chem. Soc.* 119 103-111.

INDEX

#

A

B

C

D

E

G

H

I

K

L

M

N

Q

R

T

U

T

U

V

W

X

Y

Z